Engineering and Technology Talent for Innovation and Knowledge-Based Economies

Mahmoud Abdulwahed · Mazen O. Hasna

Engineering and Technology Talent for Innovation and Knowledge-Based Economies

Competencies, Leadership, and a Roadmap for Implementation

Springer

Mahmoud Abdulwahed
Qatar University
Doha
Qatar

Mazen O. Hasna
College of Engineering
Qatar University
Doha
Qatar

ISBN 978-3-319-46438-1 ISBN 978-3-319-46439-8 (eBook)
DOI 10.1007/978-3-319-46439-8

Library of Congress Control Number: 2016952500

Printed on acid-free paper

This Springer imprint is published by Springer Nature
The registered company is Springer International Publishing AG
The registered company address is: Gewerbestrasse 11, 6330 Cham, Switzerland

Preface

This book aims to introduce systemic modelling of engineering and technology talent management for innovation and knowledge-based economies (iKBE). This includes models for engineering leadership and competencies skills, as well as frameworks for industry, academia collaboration and engineering education organizational structures incorporating innovation and knowledge-based economies development. The book also provides methodological framework as well as ready-to-use toolkits and instruments for investigation of needs of engineering and technology skills in national contexts. In addition, the book provides an empirical study, in which the innovation and iKBE (iKBEs) and leadership competencies for future engineers were investigated. A systemic framework and road map for implementation are provided methodically and are illustrated empirically. This book concludes with conceptual modelling and proposals of specific organizational structures for implementation in science-based engineering schools in order to enable developing necessary skills for future iKBE engineering graduates.

Doha, Qatar

Mahmoud Abdulwahed
Mazen O. Hasna

Acknowledgement

Sponsor
This study has been funded by Dolphin Energy Ltd. Their financial contributions for making this work are highly acknowledged.

Researchers and Contributors
The project's principal investigators would like to acknowledge the faculty and researchers' contributions throughout the process of project: Dr. Abdelmagid Hamouda, Dr. Shaligram Pokharel, Bassnt Yasser, Waled Balid, Jumana Al Madad, Kawthar Mahmoud Tawengi, and Reem Khair.

Contents

About the Authors

Dr. Mahmoud Abdulwahed holds BSc, MSc, and PhD in electrical, control and systems engineering; he completed his postgraduate studies in Germany, Sweden, and the UK. His main expertise is in Education, Innovation, and Transfer/Entrepreneurship. Mahmoud has ranked among top 25 scholars globally in the field of Engineering Education Research in last three years in terms of indexed publications in Scopus (2013, 2014, and 2015 period). Mahmoud worked in the UK before joining the College of Engineering at Qatar University in Fall 2011 as a faculty member, he was seconded from the College to the Office of Vice President and Chief Academic Officer (VP & CAO) in Spring 2013 where he worked as Adviser and then Manager of Strategic Initiatives & Innovations; his main role focused on inception, coordination, and/or management of several large scale strategic initiatives in addition to inception of innovation-based practices across campus. These included "Foundation of Excellence Initiative", "First Year Experience Implement Initiative", "Students At-Risk Initiative", "Technology Enhanced Learning Initiative", etc.; he also drove inception/approvals of several entities including the "Office of Strategic Initiatives & Innovations", the "Academic Success Unit", and the "Instructional Technology Support Unit" with the VP & CAO office, and the "Technology Innovation & Engineering Education (TIEE) Department" with the college of engineering, where he moved in Academic Year 2015/2016 to work on its implementation. Mahmoud received several fellowships, from University of Technology Sydney, University Science Malaysia, and from USA Department of State (DoS) Professional Fellows Program. He published 2 Books, and 60+ peer-reviewed conference and journal articles, and developed 40+ institutional report/framework or strategy. Mahmoud attained several industry research funds, academic recognitions, awards, and best papers distinctions. Mahmoud published on aspects related to Internet of Things (IoT), digitally enabled learning, entrepreneurship, leadership, design, ethics, and innovation and knowledge-based economies (iKBEs).

Dr. Mazen O. Hasna received the B.S. degree from Qatar University, Doha, Qatar, in 1994, the M.S. degree from the University of Southern California (USC), Los Angeles, in 1998, and the Ph.D. degree from the University of Minnesota, Twin Cities, in 2003, all in electrical engineering. In 2003, Dr. Hasna joined the Department of Electrical Engineering at Qatar University as an assistant professor. He served as the head of the Department of Electrical Engineering, the associate dean for academic affairs, and the dean of engineering between 2005 and 2012. Currently, he serves as the vice-president and chief academic officer of Qatar University. Dr. Hasna is a founding member of the IEEE section in Qatar and the Society of Qatari Engineers. He is a member of the Joint Management Committee of the Qatar Mobility Innovation Center (QMIC) and a board member of Qatar National Library (QNL).

Dr. Hasna's research interests span the general area of digital communication theory and its application to performance evaluation of wireless communication systems over fading channels. Current specific research interests include cooperative communications, ad hoc networks, cognitive radio, optical communications, and network coding. He was selected recently to the 2015 Thomson Reuters' highly cited list for his research contributions in the area of wireless communications. Dr. Hasna is active also in the area of engineering education and has published several papers in the field, and he recently edited a book about advances in engineering education in the MENA region.

Chapter 1
Engineering and Technology for Innovation- and Knowledge-Based Economies (iKBEs), Book Objectives, and Overview

1.1 Introduction and Book Scope

Several countries, whether developed or developing, are more progressively seeking transformation into innovation and knowledge-based economy (iKBE). Engineering and technology play a core role in such a transformation; hence, outcomes of engineering education and attributes of engineering graduates are crucial factors to consider. Engineering graduates are expected to be well equipped with a set of twenty-first century skills which could enable them to the future practice of iKBE. These skills and competencies need to be well-identified in the context of any country that seeks transformation into iKBE. This book aims to bridge this gap in informing policy and practice stakeholders of needed attributes as well as providing a set of tools and instruments for the investigation of needed skills in the local context in light of global advances. Furthermore, this book provides a systematic framework and well-designed road map to be developed. A case study in the context of Qatar is provided throughout the book as an illustration and empirical investigation of the developed methodologies and frameworks. The book is of a global context, and with little tailoring, the case of the Qatar investigation can be applied elsewhere using the same models and methods developed in the book.

1.1.1 Empirical Investigation Objectives

The book aims to achieve the following objectives:

- Identify attributes required by entry-level graduate engineers in the industry in three phases: current, mid-term (e.g. 2022), and long-term horizons (e.g. 2030).
- Measure the level of satisfaction with the qualifications/preparedness of current engineering graduates.

M. Abdulwahed and M.O. Hasna, *Engineering and Technology Talent for Innovation and Knowledge-Based Economies*,
DOI 10.1007/978-3-319-46439-8_1

- Address the comparative quality of students based on various factors such as gender, nationality, and institution of graduation.
- Define the importance and means of enhancing industry–academia collaboration and industry willingness for it.
- Propose recommendations for the engineering education systems to better prepare skilled graduates and overcome any potential barriers.

1.1.2 The Book's Main Contributions

The book's main contributions are as follows:

- **Literature Review**: Comprehensive literature review on competency needs of engineering talent spanned over engineering literature and non-engineering literature (iKBE, employability, graduate attributes, etc.). More than 600 studies were part of the review, making it probably the large-scale review ever conducted on this topic.
- **Theoretical/Conceptual Models**:
  - Four dimensional global model of twenty-first century talent competencies, composed of 22 skills.
  - Systematic model of competency development: a systematic framework sketching clearer vision over skills development and mechanistic relationships between two main spaces: 1—education and training and 2—workforce real-life context.
  - Systemic framework of competencies investigation, gap identification, and strategic implementation; the framework provides a basis for building upon project findings.
- **Toolkits**:
  - Quantitative and qualitative instruments as a data collection toolkit for similar investigations in other countries.
  - Systemic framework of competencies investigation also can be adopted as a systemic toolkit for similar investigations in other countries.
- **Empirical Findings**: Empirical analysis and findings of gaps and needs in the context of Qatar for the following dimensions:
  - Engineering talent competencies (currently and in the future).
  - Further perspectives on engineering education for workplace preparedness, employability, and industry–academia collaborations.
- **Propositions**:
  - Focus on excellence in technology innovation, entrepreneurship, and engineering education.
  - Aspirational engineering education systems for 2030.

- Establishment of TIEE structures.
- National technology innovation & entrepreneurship steering committee and strategic implementation road map integrating related assets in industry, academia, and government.

1.2 Role of Engineering in Innovation- and Knowledge-Based Economies

In facing today's globalized environment, the focus on maintaining or transforming into a innovation and knowledge-based economy has been significant in many countries. The Organization for Economic Co-operation and Development (OECD) defines the innovation and knowledge-based economies (iKBEs) as "*economies, which are directly based on the production, distribution and use of knowledge and information*" (Embracing Scientific and Technological Changes, nd). In view of that, iKBE is considered to play a vital role in creating wealth, improving the general well-being of citizens and fostering growth (APEC 2000). Accordingly, innovation and knowledge are found to be a necessity for creating values and outpacing wealth and ultimately strengthening the economic base of a country (Economic Importance of Engineers 2009). To facilitate the realization of knowledge, one key driver is to invest in human capital (Clausen 2004). Therefore, an advanced education system is critical for leading this transformation process, not only in motivating and preparing people to create, use, and disseminate knowledge, but also to support innovation and technology, which are at the core of iKBEs. Proper and effective engineering education represents a powerful engine that produces engineers with the required knowledge and skills for addressing future challenges and meeting the demanded growth in Qatar (Qatar National Development Strategy 2011). Because of the diverse nature of the engineering profession, engineering education should prepare graduates with adaptable skills that fit a wide range of operational, managerial, and/or technical-oriented careers ranging across multiple sectors of the whole economy (David and Foray 2002). Innovation-driven economy is considered the highest level of economic development, where countries focus on the development and use of cutting edge technology, acquiring the best talent, grooming them, and retaining them to excel in value addition through innovation within and outside the nation. The use of cutting edge technology and expertise refers to what we call the emphasis on Innovation and Knowledge Based Economy (iKBE).

The cornerstone role of technology and engineering in driving iKBEs, the industry needs, and the emergence of the engineering and technology profession has led to a significant number of studies, efforts, and investigations for identifying the needed engineering talent in the twenty-first century (NAE 2004, 2005; Jamieson et al. 2009; Mena et al. 2012; Hundley et al. 2012; Knight 2012; Spinks et al. 2006; Bourn and Neal 2008). In the Middle East, the GCC, and Qatar, an investigation similarly has not been conducted; hence, this study sheds important perspectives on bridging an empirical gap in the literature and contributing to better design of future engineering education in Qatar.

1.3 The Case of Finland

Finland at the beginning of twentieth century was a resource-based economy very similar to many developing countries nowadays. During the second half of the twentieth century, Finland transformed from an agriculture-based economy into an innovation and knowledge-based economy. This drastic transformation was a result of several governmental actions and policies coupled with advancement in education and industrial ecosystem of the country. In particular, the following factors have been crucial in the transformation of Finland's economy: 1—riding the wave of emerging industry, 2—implementation of a competitive education system, 3—systematic governance of Finland's knowledge ecosystem, 4—design and implementation of effective innovation policy, 5—monitoring and evaluation of investment impact, and 6—active connection with the global network.

One of the crucial governmental decisions that triggered rapid progress into iKBE in Finland in the 1990s and forward is to focus governmental expenditure in education, research, development, innovation, and the ICT sector in particular. This led to the emergence of the competitive ICT sector in Finland, characterized by Nokia, which formed significant GDP proportion. Following the decline of Nokia due to innovative competitors such as Apple and Samsung, Finland's ICT sector continued to grow with transformation further into software and services.

Education in Finland played a significant role in iKBE transformation. The education system in Finland has been characterized by several unique factors, in particular: 1—competent and highly educated, highly paid teachers, 2—free education system for all, 3—basic education highly comprehensive with several benefits, 4—infusion of the concept of lifelong learning, and 5—smooth emergence of education system governance from control to autonomy.

As Finland currently is moving into a second wave of iKBE, the new areas of innovation policy focus are on open innovation, disruptive innovation, co-creation, accelerators and start-ups, and fostering science and technology innovation for social and economic challenges.

1.4 The Case of Singapore

Singapore started their economic transformation in the early 1960s shortly after independence from Malaysia. The phases of economic transformation of Singapore can be classified into the following: 1—labour-based growth during the 1960s, 2—skills-based growth during the 1970s, 3—capital-based growth during the

1980s, 4—technology-based growth during the 1990s, and 5—knowledge- and innovation-based growth from 2000 and forwards. R&D expenditure in Singapore has grown from 0.5 % in the 1960s to 2.3 % of GDP in recent years. Similar to Finland, Singapore invested heavily in ICT and in the education ecosystem. In particular, Singapore's educational investment targeted technical and tertiary education, in partnership with leading international institutions. The number of scientists and engineers in Singapore has grown from 28.2 to 87.9 per 10,000 workers between 1990 and 2000. Manufacturing and services have led the economic growth of Singapore during the 1990s. In the last decade, Singapore started significant investment in the biomedical industry domain.

1.5 The Case of Qatar

1.5.1 Qatar National Vision 2030

Qatar's aspiration to become a hub of innovation and transform the country from a resource-based economy to a innovation and knowledge-based economy by 2030 is something that can be achieved through the collaboration between all parties (government, academia, industry, and services sector). Reaching this goal demands a transformation process at four main pillars as described in QNV2030; human sustainability is one of the four main pillars. The educational system is considered the base for any innovation and knowledge-based economy in the world. The human development pillar aims to have curricula and training programmes responsive to the labour market and ensure that opportunities are available to all Qatari citizens in addition to promoting and developing mindsets for lifelong learning.

As a first step for implementation, a five-year national development strategy (Qatar National Development Strategy 2011) was laid down. The NDS was the guiding document for sectorial reports in the different areas including health and education. These reports were used by different entities for developing their strategic plans and to show their support and commitment to the country vision.

Specifically from the economic prosperity angle, Qatar aims to leverage its economic status by converting its hydrocarbon-driven economy into a diversified economy that constitutes for knowledge and high-value industrial and service activities. This transformation process is expected to be done through the use of wealth generated from the available natural assets to widen investments in its infrastructure, improve its public services, develop a competitive workforce, and support both innovative and entrepreneurial initiatives whereby technology and engineering play a core vital role (OECD 1996).

1.5.2 iKBE Position of Qatar Internationally

The World Bank has developed a benchmarking tool based on a so-called knowledge assessment methodology (KAM), to classify iKBEs into an indexing scheme. According to a KAM 2012 report (SME Financing in Canada, 2002), Sweden enjoys a leading position with the highest Knowledge Economy Index (KEI) of 9.43 at worldwide level. At global level, the current status of Qatar in terms of KEI rank, which is a relative global measure, stands at 54th place with a KEI equal to 5.84. Qatar is preceded by all of the Gulf Cooperation Council (GCC) countries except for Kuwait. At the GCC level, Qatar sits at the fifth-place position headed by United Arab Emirates (6.94), Bahrain (6.90), Oman (6.14), and Saudi Arabia (5.96), which have international KEI ranks of 42nd, 43rd, 47th, and 50th, respectively.

The Global Innovation Index (GII) is another measure of iKBEs; Qatar has ranked 41st in the GII in 2013 and 50th in 2015 (GII 2015). Qatar indeed is increasingly spending on assets related to innovation since 2008, which in return builds on innovation-based country progressively; however, it seems that there are countries progressing in innovation at a faster pace, hence the lower ranking of Qatar in 2015 compared with 2013.

1.5.3 Drivers of iKBE in Qatar

There are four main dimensions that are driving Qatar's economy into a iKBE: 1—Qatar National Vision 2030, 2—World Cup 2022 and the associated mega projects, 3—instability of oil and gas prices, and 4—regional growth and competition for skills, products, and investments in the GCC region mainly led by UAE and Saudi Arabia.

Qatar is increasingly investing in knowledge development within its boundary, either through home-grown manpower or through collaborative efforts from research centres from around the world. The country's investment in research and development, education and information, and communication technology is expected to create the environment for innovation and knowledge-based economy formation. The promotion of a new economy is also supported through the establishment of international universities, science and technology parks, tax-exempt free zones, and relaxed employment recruitments in the tax free zones.

Since engineering and technology are the very heart of innovation and transformation in a innovation and knowledge-based economy, producing the next generation of engineering talent with the needed set of skills is crucial. Hence, is the importance of this investigation conducted in Qatar in light of contextual needs and global advances in engineering innovation and education.

1.6 The Importance of Leadership Skills for iKBE Development

Transition from a resource-based to innovation- and knowledge-based economy is a highly dynamic and complex process. Leaders and stakeholders driving such transformation would be significantly immersed into situations and cases that require multiple leadership attributes. Several issues such as visioning, decision-making, micro- and macroperceptions, creativity, innovation, problem-solving, systemic thinking, implementation and monitoring, driving motivation, etc., would be involved in day-to-day aspects of job functions of iKBE transformers. In particular, policy makers and engineering and technical forces are the core of transformation stakeholders. This would highlight the significance of leadership attributes for these stakeholders. Further details on leadership will be provided in subsequent chapters of the book.

1.7 Conclusions

Engineering and technology are vital cores of innovation and knowledge-based economies. Several countries worldwide are actively seeking a rapid transformation into iKBE, and this book aims to assist engineering education systems for better development of engineering curriculum and graduate attributes allied with contextual engineering talent needs. This chapter provided an overview of the book scope, objectives, and outcomes in the light of innovation- and knowledge-based economy transformation.

References

APEC. (2000). *Towards Knowledge-Based Economies in APEC*, APEC Secretariat.

Bourn, D., & Neal, I. (2008). *The global engineer: incorporating global skills within UK higher education of engineers. Engineer.*

Clausen, J. M. (2004). *The Danish Strategy, Denmark's opportunities in the global knowledge society. On behalf of Strategy Group.*

David, P. A., & Foray, D. (2002). An introduction to the economy of the knowledge society. *International Social Science Journal, 54*(171), 9–23.

Dutta, S., & Mia, I. (2011). The Global Information Technology Report 2010–2011, Transformations 2.0 (10th Anniversary Edition). From: INSEAD and World Economic Forum.

Economic Importance of Engineers. (2009). Prepared by DKM Economic Consultants Ltd. For Engineers Ireland.

Embracing Scientific and Technological Change. Shanghai Manual—*A Guide for Sustainable Urban Development in the 21st Century*, Chapter 7.

Hundley, S., Patricia Fox, Lynn G. Brown, Alan Jacobs, Catherine Didion, Daniel R. Sayre, and Hans J. Hoyer, "Attributes Of A Global Engineer: Field-Informed Perspectives,

Recommendations, And Implications," American Society for Engineering Education, no. AC 2012–4233, 2012.

Jamieson, L., Brophy, S., House, N., Harris, M., Delaurentis, D., Howell, K., Benner, A., Cekic, O., Moseir,N., Okos, M., (2009) *"Purdue's Engineer Of 2020: The Journey," in 2009 Annual Conference & Exposition; New Learning Paradigms II.*

Knight, D. B. (2012). In Search of the Engineers of 2020: An Outcome-Based Typology of Engineering Undergraduates, AC 2012-3337. In *Proceedings of the 119th Annual Conference of the American Society for Engineering Education,*. San Antonio, TX.,.

Mena I.B., Zappe, S. E., and Litzinger, T. A., *(2012), "Preparing the Engineer of 2020: Analysis of Alumni Data," in 2012 ASEE Annual Conference, American Society for Engineering Education.*

NAE. (2004). *The Engineer of 2020: Visions of Engineering in the New Century.*

NAE. (2005). Educating the engineer of 2020: Adapting engineering education to the new century. *IEEE Engineering Management Review*, 1–208. doi:10.1109/EMR.2009.4804343

OECD. (1996). *The Knowledge-Based Economy*. Paris: Organization for Economic Co-Operation and Development.

Qatar General Secretariat for Development Planning (2011). Qatar National Development Strategy (QNDS) 2011 ~ 2016.

Small and Medium Enterprise (SME) Financing in Canada, Government of Canada Report (2002). Appendix A, page 100.

Spinks, N., Silburn, N., & Birchall, D. (2006). *Educating Engineers for the 21st Century: The Industry View. Engineering*. UK: Greenlands.

Chapter 2
Literature Review of iKBE Competencies and Systemic Modelling

2.1 Introduction

Historically, the main skills engineers need to have are technical disciplinary content knowledge and technical competency; however, concerns of needs of engineers for a variety of non-technical skills can be traced to the early 1980s (Wearne 1984). Engineering in particular and other relevant STEM (Science, Technology, Engineering and Mathematics) fields in general are the most critical sciences for a nation to consider while seeking for or maintaining a Innovation and Knowledge-Based Economy status (NAE 2005, 2004; NRC 2007; Pinelli and Haynie 2010). In this context, the talent and skills requirements of current and future engineers as well as necessary relevant reforms in engineering education curriculum have become a hot topic of investigation during the last 10–15 years. Numerous studies and policy reports have been published mainly in developed countries or emergent economies, such as USA (Duderstadt 2008a), Australia (Rabl and Hillmer 2012), UK (Spinks et al. 2006), Canada (Chan and Fishbein 2009), Malaysia (Abdullah et al. 2007), India (Mishra 2010), and Thailand (Sunthonkanokpong 2011). The majority of studies have investigated needed talent in engineering in general; however, some research has had a more microscopic level, focusing on the required skills in a specific domain of engineering. For instance, studies have been published on skills required in transportation engineering (Skills and They 2009), mechanical engineering (2028 Vision for Mechanical Engineering: A report of the Global Summit on the Future of Mechanical Engineering 2008; Danielson 2011), engineering management (Dudman and Wearne 2003), and civil engineering (ASCE 2007). These studies continue to emphasize that technical content knowledge and competencies are essential for any engineer; however, in addition to being well grounded in mathematics and science, twenty-first-century engineers should be well shaped in broader knowledge base and diverse personal/interpersonal key skills. Such attributes and skills may include the following: teamwork, communication,

M. Abdulwahed and M.O. Hasna, *Engineering and Technology Talent for Innovation and Knowledge-Based Economies*,
DOI 10.1007/978-3-319-46439-8_2

inter/multidisciplinary knowledge, analytical thinking, ingenuity, creativity, technological innovation, business and management skills, leadership, ethics, professionalism, as well as understand work strategies (Anderson et al. 2005; Sheppard et al. 2004; Swearengen and Barnes 2002; Shuman et al. 2005).

This chapter aims to provide a comprehensive literature review on needed skills for graduate and future engineers, in the context of generic skills and employability needs in knowledge-based societies and economies. The review spans over a diverse range of relevant studies, reports, books, conference proceedings, and journal articles. The review findings and contributions can be coined in three main folds: 1—a conceptual ontological framework of engineering skills in the context of generic—and employability—skills of KBSs and iKBEs citizens and workforce, 2—the development of a model of driving forces for focus on shift of needed skills and attributes of engineers, and 3—a global set of 22 mutual common skills, but different in topology, between engineering and generic competencies.

2.2 Drivers of Focus on Skills Development in Engineering Education

There have been many drivers for the emergence of research on skills and talent of graduate engineers, and the same drivers have led to identifying the needs of engineering graduates and skills that meet significantly wider needs than ever. These drivers are mainly the following:

- Satisfying industrial needs (Spinks et al. 2006).
- The emerging roles and disciplines of engineering sciences (Tryggvason and Apelian 2006).
- Adapting to highly advanced and complex working environment.
- Contextualizing the generic engineering graduates' attributes in the light of the needs of knowledge-based societies and economies (Male and Chapman 2005).

These factors have been pressing on deploying changes in engineering education systems to meet the emerging needs. The drivers are visually represented in the model-of-four-forces (eng-skills-MoFF) shown in Fig. 2.1 and are detailed in the subsections below.

2.2.1 Driver 1: Bridging the Gap with Industry Needs

The gap between the graduates' actual and industry-desired attributes has been extensively investigated and recognized through numerous studies, mostly surveying engineering employers about their satisfaction with the fresh graduate skills and competencies. The majority of the studies have reported a deficiency in

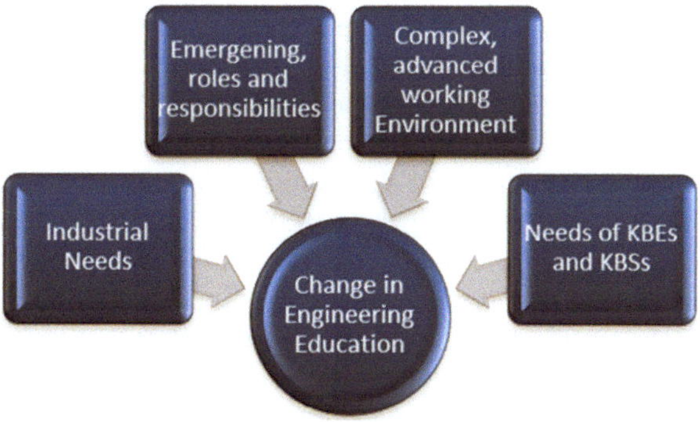

Fig. 2.1 The model-of-four-driving-forces for focus on engineering skills (eng-skills-MoFDF)

graduate attributes (Kirkpatrick 2011; Rabl and Hillmer 2012; Patil and Codner 2008; Zaharim et al. 2010; Spinks et al. 2006) mainly in soft skills such as communication skills, leadership, social and ethical skills. Nowadays, industrial bodies are looking for graduates who are equipped not only with technical but also with many other soft skills (NSB 2007), and for some employers soft skills are even more valuable and appreciated (Khair et al. 2013).

The reason behind this gap is the imbalanced focus of the engineering education on teaching engineering knowledge and technical skills at the expense of enforcing soft skills in the students (Childs and Gibson 2010). As a response, the US and Australian engineering education systems have shifted over the last 15 years towards an outcome-based approach to produce graduates with necessary skills for the industry market (Walther and Radcliffe 2007).

The gap between education and practice of engineering can widen in the future (NAE 2005). Especially in the GCC and Middle East, the gap is higher due to the rapid development and growth dynamics in the market and industry that have not been coupled with adequate educational advancements to keep an alignment between both. Hence, prompt actions need to be taken to advance the engineering education system, to better prepare the graduates for practice, and to satisfy the need of industry.

2.2.2 *Driver 2: The Ever-Evolving Emergences, Roles, and Responsibilities of the Engineering Profession*

It is argued that the definition of engineering should be expanded to cover the emerging trends and roles of modern and future engineering (Tech 2011). Felder (1998) indicates that an engineer with technical knowledge only is no longer sufficient to solve the new complex and interdisciplinary problems. Twenty-first-century engineers are expected to invent solutions to "grand challenges" which are global in nature, such as poverty, health, environmental concerns, housing, food preservation, and political instability (Klein-gardner 2011; NSB

2007; King 2008). Furthermore, they need to understand global markets and business practices; understand global ethical, safety, and security standards; have cultural, social, and political awareness (Chang et al. 2009; Patil and Codner 2008; Klein-gardner 2011), and pay attention to economic, environmental, political, and social constrains while developing solutions for their future problems (Calfee 2009; Mishra 2010). Chang et al. (2009) have stressed that global engineers "*need to become more aware of and responsible about their societal problems, investigate them, and put innovative solutions*".

Engineering/STEM graduates of twenty-first century are not only demanded for technical and scientific careers, but also started to play essential roles in non-traditional fields such as business, finance, management, policy, and social studies.

2.2.3 Driver 3: Adapting to Highly Advanced and Complex Working Environment

There is a huge international expansion of today's market, companies, and supply chains and their tendency to work on projects with collaborations from different countries (Mishra 2010; Parkinson 2009). In this open world of advanced technology, open trade, and multinational companies, engineering graduates are expected to have global competence (Parkinson 2009) and to be able to solve various types of problems under various constrains and in different circumstances (Kastenberg et al. 2006). Such working environments involve team members of various nationalities, cultures, and languages. (Mishra 2010; Duderstadt 2010; Klein-gardner 2011) Consequently for these fairs and projects to be successful, they require their members to have not only good technical but also soft skills, such as communication and teamwork skills (Mishra 2010).

Dealing with advanced technology and complex systems in engineering problems requires highly interdisciplinary engineering teams who have a broad mental span instead of narrow practice within traditional disciplines (Duderstadt 2010; NAE 2004; Rabl and Hillmer 2012). Engineering graduates will need to efficiently interact and collaborate with other engineers including graduate of emerging engineering disciplines such as bio and genetic engineering, medical engineering, financial engineering, nanoengineering, and ecoengineering. (Agogino 2005; NSB 2007) in addition to experts from non-engineering professions (NAE 2004; Educate to Innovate 2009).

Alongside the change in the market, engineering curricula have to change and incorporate subjects of globalization, multiplicity, world cultures and languages, communication, leadership, and ethics, environmental regulations, and sustainable approaches in design as core elements (Jamieson et al. 2009; Mishra 2010). Some well-known universities, such as Purdue, have developed their engineering programme into a "global engineering programme" with the goals of developing and improving the students' soft skills allowing their graduates to standout in culturally diverse teams (Jamieson et al. 2009).

2.2.4 *Driver 4: Compliance with the Generic Needs of KBSs and iKBEs*

One major contributor to the transformative shift of considerations of needed skills of engineering graduates is the requirements of knowledge-based societies (KBS) and economies (iKBE) (Male and Chapman 2005). Tryggvason and Apelian (2006) refer to the emergence of engineers' roles during the last two centuries as follows: 1—the professional engineer: nineteenth and first half of twentieth century; 2—the scientific engineer: second half of the twentieth century; and 3—the entrepreneurial engineer: twenty-first century. Entrepreneurial engineers are engines of innovations that are driving high value-adding economic growth of twenty-first-century Innovation and Knowledge Based Economies and nations. iKBE or KBS generic skills and competencies such as communication, lifelong learning, and adaptability have also become frequently cited in the literature on required skills and attributes for engineers (NAE 2004, 2005; RAE 2007). It is worth mentioning that certain core engineering skills such as problem-solving and analytical thinking became essential even for those who would study and work in non-engineering disciplines; such core engineering skills are becoming frequently cited as necessary generic skills for the workforce and citizens of KBSs and iKBEs.

2.2.5 *Literature Review Methodology*

This is a literature review aimed at identifying related studies on competencies and skills needed of twenty-first-century engineers. A number of relevant keywords were utilized in the search of related literature in Google Scholar, as well as research databases such as ScienceDirect and IEEE Explore. Keywords were included, for example "future engineers", "engineers of 2020", "engineering employability skills and attributes", "graduates' attributes", "twenty-first-century skills", "global engineer", "generic skills", and "graduate employability".

The focus of the scan was on identification of key current and futures engineers' skills set, employability skills, skills set requirements of citizens of iKBEs and KBSs, different research methodologies that have been followed, and any developed frameworks and global approaches utilized to integrate these skills in education and training.

More than 500 studies, reports, and research papers were found. After initial filtering, about 190 of the found documents have been selected for full review. The reviewed articles included about 150 research papers and 40 reports. The majority of studies and reports had a specific national context, such as

- USA-focused studies (27 documents): (Casner-Lotto and Barrington 2006; Chang et al. 2009; Co-sponsors et al. 2008; Danielson 2011; Duderstadt 2011; Finegold and Notabartolo 2010; HRA 2013; Hundley 2012; IFEES n.d. 2012; Jamieson et al. 2009; Knight 2012; Mena et al. 2012; Michigan 2009; NAE 2004, 2005, 2011; NSB 2007; P21 2008; Palmer et al. 2011; Parkinson 2009; Patil and Codner 2008; Rajala 2012; SCANS 1991; Shuman et al. 2000a; Terenzini and Lattuca 2011; Tippins and Hilton 2010),
- Australia-focused studies (14 documents): (ACER 2002; Bowman 2010; Dawe 2002; DEST 2006, 2007; Gibb 2004; Kerr 2010; SA Male 2010a; MSA 2012; NCVER 2003; Nguyen 1998; Rabl and Hillmer 2012; RAY Townsend and Waterhouse 2008; Wheelahan et al. 2011),
- UK-focused studies (11 documents): (Allan and Chisholm 2009; CBI 2007, 2009, 2012; Edinburgh 2012; IEA 2009; Markes 2006; Martin et al. 2008; PyeTait 2011; Spinks et al. 2006; UKCES 2009),
- Malaysia-focused studies (6 documents): (NHEAP 2007; Nor et al. 2008; UNESCO 2012; Zaharim et al. 2009; Zaharim et al. 2010),
- Canada-focused studies (3 documents): (Canada 2000; Chan and Fishbein 2009; NESP 2010),
- Hong Kong-focused studies (2 documents): (Fung et al. 2006; Noakes 2004),
- India-focused studies (2 documents): (Mishra 2010; Saravanan 2006),
- Ireland-focused studies (2 documents): (Childs and Gibson 2010; Statz 2003),
- Scotland [1 study (Hounsell 2011)],
- Netherlands [1 study (REFLEX 2007)]
- Japan [1 study (Nguyen et al. 2005)],
- Indonesia [1 study (UNESCO 2012)],
- Philippine [1 study (Llanes 2008)],
- Singapore [1 study (WDA 2009)], and
- Thailand [1 study (Sunthonkanokpong 2011)].

Some of the other reviewed studies have had an international or continental scope (Continental 2006; ERT 1995; SHAH 2009). No studies were found from the Gulf Corporation Council (GCC), Middle East, or North Africa.

All reported skills in the reviewed studies have been synthesized. A content analysis approach was utilized for systematically categorizing them and also for generating a generic definition for each category. Content analysis is an observational research technique that can be used to systematically evaluate any content in order to identify the scripts in terms of various categories (Li and Cavusgil 1995; Kolbe and Burnett 1991; Holsti 1969). Definitions, terminologies, and types of skills and attributes were synthesized, and a relevant conceptual framework has been developed.

2.3 Engineering Education Accreditation Systems and Engineering Competencies

Undergraduate education plays a significant role in the development of engineering and employability skills of graduates. It shapes and refines these skills through students' involvement in various projects, assignments, lectures, etc. Higher education institutions usually aim to satisfy the accreditation requirement of different national and international accreditation bodies to achieve and prove high-quality educations experiences for its students. Over the last 2 decades, and in the light of recognizing the importance of students' learning outcome, a dramatic shift in the accreditation requirements has occurred from being input-oriented to output-oriented criteria. Thus, the focus is now on what the students are competent in rather than on what resources have been employed in teaching them. Satisfying these criteria and being accredited guarantee that the students of such programmes attain certain outcomes including a set of soft skills. In this study, the review of various engineering education accreditation systems has been conducted and a set of competencies required were extracted in the process of developing competency needs of twenty-first-century engineers. Engineering accreditation systems from USA, UK, Europe, Australia, Japan, Singapore, and Malaysia were reviewed. Appendix A provides details on the engineering attributes these systems require.

2.4 Synthesis of Global Set of Skills

The synthesis of a global set of skills of twenty-first-century engineers has been achieved in four stages:

1. Identifications of various terminologies in regard to skills or competencies,
2. Skills extraction from the literature on skills and competencies (engineering and non-engineering),
3. Model generation, and
4. Counting and quantifying.

Further details on each stage are provided in the next subsections.

2.4.1 Definitions: Skills, Attributes, Competencies, and Others

2.4.1.1 Generic Literature (Non-engineering)

There has been a wide range of terminologies and definitions that have been utilized to describe and define the different terms of skills and attributes. The term "skills" is normally utilized to refer to the ability of performing a task. The term "attribute" refers to a characteristic or feature of something and it might be part of the personality nature or might be developed through life experience. Both skills and attributes constitute key elements for employability. They both refer to the ability to apply content knowledge in a practical way in order to accomplish a mission or a task. Knowledge is defined as the interaction between the capacity (intelligence) and opportunity (situation) to learn something, and it includes theory and concepts. The term ability refers to basic competence (skills, knowledge, and attitudes). The term competence has also been defined by the Organization for Economic Cooperation Development (OECD) as "*the ability to successfully meet complex demands in a particular context*" (OECD 2002). It is indicated that the development of competence issue is influenced by a number of factors that are as follows: (a) ability, (b) knowledge, (c) understanding, (d) skill, (e) action, (f) experience, and (g) motivation. The OECD classifies competencies into the following: 1—"key competencies" and 2—"specific domain competencies".

Key competencies are defined considering those "important across multiple areas of life and contribute to an overall successful life and a well-functioning society" (OECD 2002), while specific domain competencies are those which "do not apply across multiple areas of life and are not necessary for everyone or are irrelevant to the betterment of individual and societal life" (OECD 2002). The term "key competencies" coined by the OECD is similarly referred in other literature by the term "generic skills" or "generic competencies" (Hager and Holland 2006; Male and Chapman 2005). Internationally, generic skills are known by a number of different terms from one country to other (NCVER 2003; UNESCO 2012) such as basic skills, necessary skills, and workplace know-how (USA). Generic skills can be referred as core skills, key skills, common skills (UK); key competencies, employability skills, generic skills (Australia); employability skills (Canada); critical enabling skills (Singapore); transferable skills (France); key qualifications (Germany); transdisciplinary goals (Switzerland); and essential skills (New Zealand).

According to Hager and Holland (2006), the term "generic skills" has been widely utilized recently across education systems for the adjustment of outcomes of graduates, where they are often called "graduate attributes" or "graduate qualities". The term "generic skills" refers to a wide range of qualities and capabilities, which can be applied in different aspects of life, work, and academia. Hence, generic skills include both graduate skills (and graduate attributes) and employability skills. Graduate attributes are defined as "*the skills, knowledge, and abilities of university graduates, beyond disciplinary content knowledge, which are applicable to a range of contexts*" (Barrie 2007). Clearly, graduate attributes are more than just skills and competencies for employment; they are life skills as well. Employability skills are sometimes mapped into a number of terms such as key skills, core skills, life skills, generic skills, essential skills, key competencies, enterprise skills, necessary skills, and transferable skills (DEST 2007). Employability skills are the bridge between academia and work and have been defined in literature as "*a set of achievements—skills, understandings, and personal attributes—that make graduates more likely to gain employment and be successful in their chosen occupations, which benefits themselves, the workforce, the community, and the economy*" (Yorke and Knight 2006).

2.4.1.2 Engineering Literature

Skills identified in the engineering context literature are normally divided into two categories: the "technical" or "hard" skills and "non-technical" or "soft" skills; both technical and non-technical are complementary skills for engineers (NAE 2004, 2005). The non-technical skills are considered important to deliver the technical skills in a proper way (NAE 2004, 2005). The non-technical skills are also called a variety of other different terms such as soft skills, twenty-first-century skills, professional skills, foundation skills, graduates' skills, functional skills, and graduate qualities (NCVER 2003). Engineering skills can be also divided into the global and the local.

2.4.2 Skills or Competencies

In our understanding, attributes, skills, and competencies have been interchangeably utilized across the literature; however, we may refer to competency as a skill with higher level in performance. In this book, the terms will be used interchangeably unless otherwise explicitly stated.

2.4.3 Skills Extraction

A synthesis of skills was performed taking into consideration two main dimensions:

- DIM1—Skills that are required for current and future engineering graduates (extracted from engineering-related literature) and
- DIM2—Generic graduate skills normally in a KBS or a iKBE context (extracted from general literature, not engineering related)

Dimension 1 focusing on engineering skills is coined with **engineering skills**, while Dimension 2 is coined with **general skills**. In this stage, the retrieved papers from the literature were reviewed and all the skills mentioned were extracted and tabulated under one of these two dimensions. The first list included skills from studies on future engineers, engineering graduates, global engineers attributes, and engineering accreditation systems; while the second list included skills from studies on employability, twenty-first-century, generic skills, and graduates' attributes. For each dimension, a list of about 200 skills has been extracted from the literature.

A total of 37 references have been utilized in synthesizing the list in the engineering skills dimension, while a total of 34 references have been utilized in synthesizing the list in the general skills dimension.

2.5 Global Competency Framework and Model Development

In our review of different set of literatures spanned over employability, graduate attributes, twenty-first-century skills, Innovation and Knowledge Based Economy competencies, engineering skills, etc., it was apparent that both engineering literature and non-engineering literature share global set of competencies that can be themed into couple of dimensions and categories. Content analysis led to categorize the skills found into a set of 22 items of global skills under 4 main dimensions.

1- **Dimension I—Core Knowledge and Practice**, which includes the following competencies: 1—science knowledge (math, physics, and science fundamentals); 2—disciplinary fundamentals, 3—interdisciplinary fundamentals, 4—multidisciplinary knowledge, 5—practical experience, and 6—ICT skills.
2- **Dimension II—Cognition, Mental, and Thinking**, which includes the following competencies: 1—lifelong learning, 2—problem-solving, 3—decision-making, 4—analytical thinking, 5—systems thinking, 6—critical thinking, 7—creative and Innovation, and 8—design.
3- **Dimension III—Professional and Interpersonal**, which includes the following competencies: 1—professionalism, 2—ethics and responsibility, 3—adaptability, 4—communications, 5—teamwork, and 6—foreign languages.
4- **Dimension IV—Business and Management**, which includes the following competencies: management, leadership, and entrepreneurship.

2.5.1 The Pyramid of Global Competencies

The four dimensions described earlier in the previous subsection are categorized into a pyramid model composed of four main levels: 1—generic, 2—domain specific, 3—subdomain specific, and 4—functional specific; see Fig. 2.2.

Level 1—Generic: The Dimensions II, III, and IV are global for twenty-first-century human talent; they can be needed in most disciplines and professions, and they are also considered as competencies for life, personal, and interpersonal relations. Dimension I in an abstract manner is generically applicable to all professions, but it functions as a differentiating dimension. The type of competencies in Dimension I will be different depending on the domain. This is what we call Level 1 in the competencies pyramid model shown in Fig. 2.2.

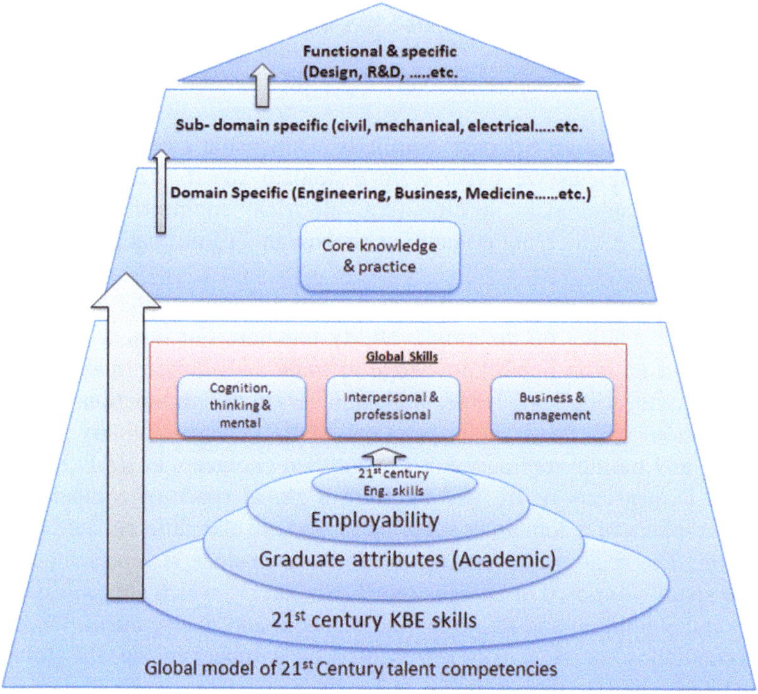

Fig. 2.2 Ontological conceptual framework of the relation among skills (from generic to specific)

Level 2—Domain Specific: Dimension I is a differentiating set of competencies depending on the discipline. For instance, in engineering the following competencies will be applicable for Dimension I: 1—sciences knowledge (math, physics, and science fundamentals), 2—disciplinary engineering fundamentals, 3—interdisciplinary engineering fundamentals, 4—multidisciplinary knowledge, 5—practical skills, and 6—ICT skills. In another field such as business, for instance, Dimension I competencies will be as follows: 1—sciences knowledge (math, physics, and science fundamentals), 2—disciplinary business fundamentals, 3—interdisciplinary business fundamentals, 4—multidisciplinary knowledge, 5—practical skills, and 6—ICT skills. Taking a discipline in engineering (Eng) and in business (Buis.) such as electrical engineering (Eng.) and accounting (Buis.), Dimension I competencies will be differentiating in these sets. For instance, practical skills in electrical engineering will be to a significant extent different from practical skills in accounting; science knowledge in engineering will be to a significant extent different from math and sciences knowledge in accounting. Electrical engineers will need to know interdisciplinary knowledge of other engineering disciplines such as mechanical and computer engineering, while accountants will need to know interdisciplinary knowledge of other business disciplines such as management or information systems. Electrical engineers will need to know apart from conventional software such as MS word, knowledge about engineering software tools such as MATLAB or CAD design, while accountants similarly need to know about MS word, but also other specific domain software systems for accounting.

Level 3—SubDomain Specific: Similarly, Dimension I competencies will get further differentiation as we move from a domain specific (e.g. engineering or business) into a subdomain specific (e.g. electrical engineering or mechanical engineering in the engineering domain, or accountant or information systems in the business domain).

Level 4—Functional Specific: This level will differentiate the competencies of Dimension 1 depending on the employability function. For instance, engineers in R&D will need more in-depth knowledge of math and science fundamentals, and disciplinary engineering knowledge, than engineers in design function. Engineers in design function will need more practical skills, interdisciplinary engineering knowledge, and multidisciplinary knowledge than engineers in R&D functions.

Overall, Dimensions II, III, and IV form a global space of competencies that operate in a sphere of a domain or subdomain, and are also differentiated depending on function. The relationship among the four dimensions is represented visually with a 3D space composed of Dimensions II, III, and IV, which are professional and interpersonal, cognition and thinking, and business and management. Dimension I (the differentiating dimension) is represented by a sphere in the 3D global competencies space; this is shown in Fig. 2.3.

The relationship among the different terminologies and concepts of skills reviewed in various literature has been systematically modelled and are shown in an ontological framework in Fig. 2.2. The analysis has shown that all of the 22 skills

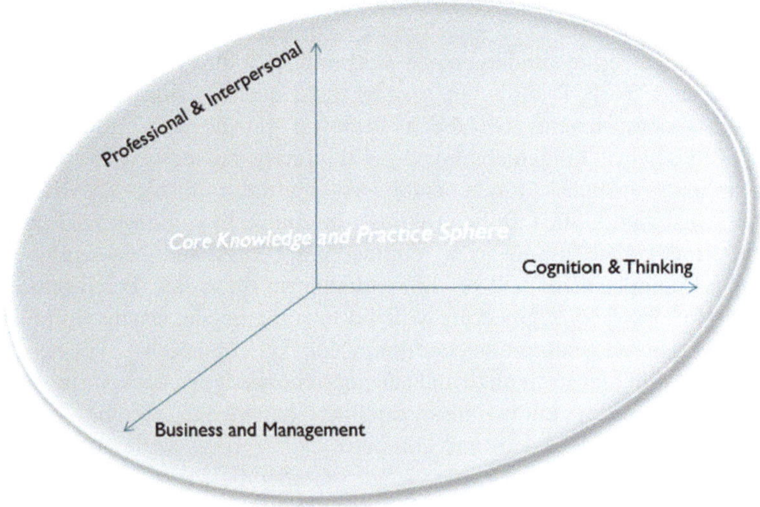

Fig. 2.3 Twenty-first-century talent competencies

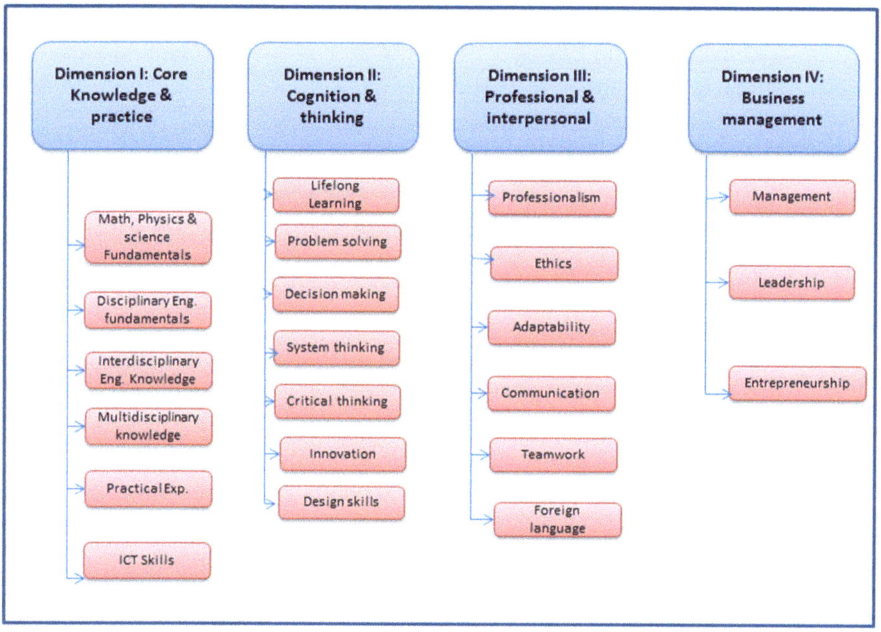

Fig. 2.4 The global set of twenty-first-century skills

are common between the two aimed dimensions, but topological differences in terms of emphasis have arisen; this will be highlighted in further details in this section. The global set of 22 skills are further defined and detailed later in Appendix B; Fig. 2.4 shows the developed model.

2.5.2 Counting and Quantifying

At this final stage, the total number of papers for each set of skills was calculated as a percentage of the total number of references for each of the two domains. This has been done to quantify the emphasis on each set of skill at each of the **engineering skills** and the **general skills** (or what we call the twenty-first-century talent competencies) literatures. Further comparisons, implications, and elaborations are given in the following section, see Fig. 2.4.

2.5.3 Discussion

The review indicated that the most highly emphasized 6 skills in the literature for the **engineering skills** are as follows: 1—communication skills, 2—business and

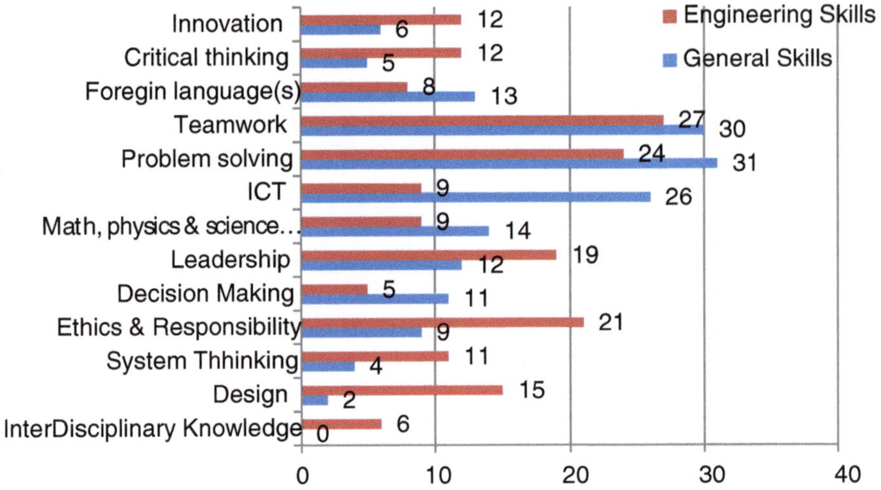

Fig. 2.5 Skill sets that have more difference in emphasis comparing both general skills and engineering skills literatures

management 3—teamwork skills, 4—problem-solving, 5—lifelong learning, and 6—ethics and responsibility. While for the **general skills** there were 1—communication skills, 2—problem-solving, 3—teamwork, 4—ICT experience, and 5—business and management. Significant gap of emphasis between the ENGD and the **general skills** is found in 12 skills. A significant difference in emphasis (i.e. ≥ 15 %) was found for 12 of the skills illustrated in Fig. 2.5. The greatest difference in emphasis was noted for ICT skills, design, ethics and responsibility, problem-solving, and multidisciplinary knowledge when comparing the generic to the engineering skills literature.

The greater emphasis on system design and thinking skills in engineering references could be attributed to the nature of the profession itself, which requires more of these skills as compared to other professions. Practical experience being more emphasized for engineering indicates the greater need for engineering graduates to be prepared for the huge complexity of the twenty-first-century working environment of engineers. Surprisingly, there is greater emphasis on ICT experience and problem-solving skills as generic skills rather than engineering skills. However, it could be hypothesized that engineering graduates are more familiar with the new technology as compared to graduates from other disciplines, such as medicine and business, requiring less emphasis in the literature. The same could apply for decision-making skills because the engineering profession involves decision-making by its nature (mainly technical, but many times non-technical too). Also, engineers during their studies have to take numerous decisions while working on projects, engineering design courses, or even complex open-ended problem-solving. The higher emphasis on ethics and responsibility of engineers shows the greater than ever need to integrate ethics into engineering education.

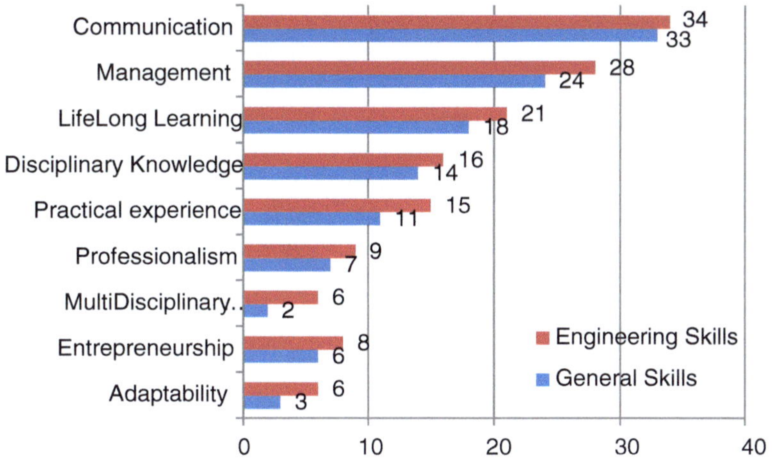

Fig. 2.6 Skill sets that have less difference in emphasis among general skills and engineering skills literatures

Ethics are vital to be considered while designing, deploying, and servicing engineering solutions. Foreign language skills have been closely emphasized as both engineering and generic skills, which can reflect the global multinational multicultural working environment of today's world. Communications skills were among the most highly emphasized set of skills from literatures. Adding to that the high emphasis on teamwork, this can be attributed to the team-based practice of many professions. Being a lifelong learner is critical not only for engineers but also for other disciplines in the world of rapidly expanding knowledge (see Fig. 2.6).

2.6 Systemic Model of Competency Development

2.6.1 Ontological Relations and Mechanics

In this section, a systemic framework for the ontological relations among the different definitions previously is developed. The main aim is to clarify the confusion among the different concepts and definitions utilized, and also to sketch a clearer visual model on the mechanistic relationship of skills development in two main spaces: 1—the education and training system, and 2—the workplace and/or authentic real-life contexts. From the previous definitions outlined earlier, one can conclude that attaining a competency passes through the following process that is spanned over two spaces: **Space 1**—the education and/or training system: *1—help to acquire a preliminary content knowledge either by training or through an education system; 2—facilitate blending of the knowledge with an intrinsic attribute and charter; 3—provide opportunities to apply the blended knowledge in an* *application context. These efforts will help to develop preliminary skills (can be called a graduate attribute), which can lead to a higher order intrinsic attributes or character.* **Space 2**—the workplace and/or authentic real-life context: *the context implies that the conditions are larger in space, require inputs from more than one discipline and require a lengthier planning and development period than those dealt with in Space 1. Therefore, the system should help to 1—acquire a needed contextual content knowledge; 2—facilitate blending of this knowledge with a higher order intrinsic attribute and charter; 3—provide exposure and opportunities to apply the knowledge on complexity in one or multiple application contexts, which can develop a higher order skill, or a competency, and would lead to a higher order intrinsic attribute or character.* The process in the second space might require an iterative phase with the first space that is iterating back to the fundamental education system to understand the recent developments in the subject area or to sharpen the skills to a higher degree. In either case, iteration becomes the key to prepare the skills to tackle the challenges of a iKBE.

2.6.2 Hypothetical Implications of the Model

The sketched systemic model of competency development shown in Fig. 2.7 may lead to a number of hypothetical implications. These hypotheses govern the relationship and reciprocal continuity between the two main spaces a skilled human resource in a iKBE goes through.

Hypothetical Implication 1—Continuity of Content Knowledge between Spaces: *Complete distinction of content knowledge between Space 1 and Space 2 may result in delay or lower order of competency development; it may even require iteration through a new education/training system. Hence, educational/training systems should align their provisioned content knowledge with the needs of the workplace and real-life contexts their graduates are expected to function.*

Hypothetical Implication 2—Continuity of Contextual Application: *Higher resemblance of contextual application in Spaces 1 and 2 would lead to higher alignment in the development of attributes, skills, and competencies. Hence, educational/training systems should align the context by which their content knowledge is provision to become resemblance (as far as possible) to those contexts normally faced in the workplace and real-life situations their graduates are expected to function.*

Hypothetical Implication 3—Emphasize on Application of Content Knowledge in Space 1: *Traditional education systems tend to emphasize heavily on content knowledge provision, paying less attention to importance of applications for skills development. Theories, such as constructivism and experiential learning, in the learning sciences criticize such approaches, and emphasize the importance of*

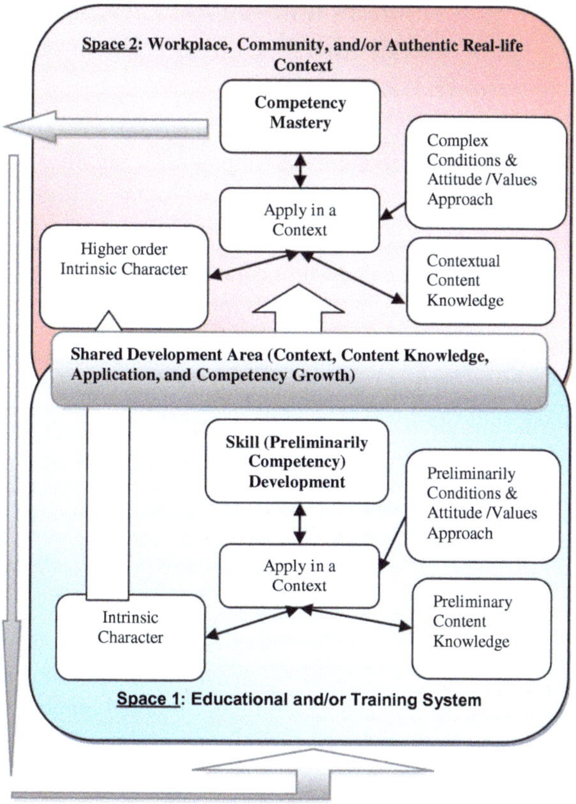

Fig. 2.7 Semantic systemic model of mechanistic and ontological relationships between content knowledge, skills, attributes, and competencies

applications for meaningful learning and skills attainment. Hence, educational/training systems that are heavily theory oriented with lack of experiential and active learning approaches for content knowledge applications may lead to poor development of graduate attributes needed for the workplace. Having modelled the dynamics and systematic development of competencies, the next subsection provides a view of needed competencies of twenty-first-century engineers in iKBEs and KBSs.

The model implies several theoretical assertions for education systems that are effective and connected to the workplace and twenty-first-century graduates. In the next section, we provide a synthesis of recommendations collected from stakeholders and from the literature on improving the engineering education system to meet twenty-first-century requirements.

2.7 Stakeholders and Global Literature Recommendations for Twenty-First-Century Engineering Education for iKBE

Throughout this book's development, a significant number of recommendations have been attained and found through one of the following roots:

- Emerged from involved stakeholders in the study (industry: managers, trainers, HR, and practicing engineers; academia: faculty and senior students) and
- Synthesized from the reviewed literature globally

These recommendations have been summarized and categorized in the subsections below. Those recommendations that emerged from stakeholders in Qatar are cited with QS as an acronym of Qatar stakeholders, and the rest are cited via the reference they have been found in.

2.7.1 Engineering Practice and Industry: Academia Linkage

The theme of engineering practice in engineering curriculum and better linkage with industry was one of the main observed emphases of interviewed stakeholders. Interviewees recommended introducing interventions for equipping students with practical skills needed for employer in advance of graduations to enhance job productivity, to reduce training programs' duration in industry, and to enhance employability for non-nationals (QS). Furthermore, all groups called for increasing the proportion of practical training and internships, even beginning with freshman year (QS); to further enable this, some have recommended developing a section or organizational structure in large companies dedicated for internships and industry–academia linkage (QS). Similarly, faculty reported that the absence of such effective structure on the academia side is one of the major barriers of enhancing industry–academia linkages. Several interviewees called to host instructors from industry in engineering courses and to include more feedback from industry in curriculum design (QS); similar calls have been recommended in international studies of Nor et al. (2008) and RAE (2007).

2.7.2 Innovation, Design, and Entrepreneurship

The theme of innovation, design, and entrepreneurship has been one of the other emerging focuses of stakeholders' recommendations, similarly with recent trends internationally indeed. Several stakeholders recommend establishing structured programs that encourage innovation and design skills in order to help to have iKBE industries [QS]. Design-based engineering education is increasingly embraced as a new philosophy; one of the most influential policy reports by the National Academy of Engineering (USA) recommend to include hands-on design projects from freshman year and in all professional years (NAE 2005). Furthermore, capstone designs were strongly recommended to be derived from real-world and industrial problems (NAE 2005). More emphasis on entrepreneurship and technology transfer in engineering education was highly recommended by interviewees as a major mean of transferring innovations in design into added value (whether economic or social) outcomes (QS).

2.7.3 Pedagogies and Engineering Education Research

Engineering education research was one of the themes that was highly recommended by stakeholders from all targeted groups. Focusing on design- and project-based approaches with less written exams, in particular in junior and senior engineering education years, was highly recommended by interviewees (QS). Many calls encourage faculty to move from traditional teaching into more proactive approaches such as problem-based learning and practical learning (Nor et al. 2008). National Academy of Engineering (USA) recommend that engineering education institutions should encourage pedagogical research among its faculty as valued and rewarded activity aiming to understand students' learning and develop best methodologies of teaching (NAE 2005).

2.7.4 Curriculum, Training, and Programmes

Engineering education programs as a theme was highly emphasized by interviewees for better enhancement. Engineering education curriculum (including initiating new programmes/courses) should be designed in an adaptive manner emphasizing in particular on management and leadership development in engineering education curriculum so that it can cope with future needs and trends of the country flexibly [QS] introducing requirements on continuous education for professional engineers to mandate frequent update of knowledge and skills after graduation and deployment in the workplace [QS]. Once new curricula or educational strategies have been implemented, continuous monitoring and reflection are strongly encouraged (Shuman et al. 2000b).

2.7.5 Engineering Talent and Competency Development

Engineering skills and competencies were highly demanded by stakeholders for better readiness to workplace. Competencies development as a well-rounded approach for enabling development of soft skills for engineering students using curricular and extra-curricular approaches was highly emphasized. (QS) (Bourn and Neal 2008). Making orientations to introduce all the majors to ensure students chose the major/career that suits them was one of the approaches recommended (QS). Developing an engineering career service in the university to assist in employability matters of fresh graduate engineers was also highlighted [QS].

2.8 The Interrelation Between Twenty-First-Century Engineering Competencies and Leadership Attributes

Throughout the review, a correlation has been identified by the global set of twenty-first-century iKBE attributes and leadership attributes. Leadership most of the times is summed into one term which is "*leadership*"; however, a deeper look on the literature of leadership space reveals a significant amount of attributes. In this book, a more critical view of leadership attributes and models is provided and has been interlinked with the global literature on twenty-first-century iKBE skills. For instance, twenty-first-century skills such as communications, teamwork, and problem-solving are core leadership attributes that are found in almost any leadership models. Furthermore, the engineering education community has been progressively developing various programmes and interventions in the engineering education curricula in order to equip future engineering graduates with leadership skills. Further details on this subject are provided in the next chapter.

2.9 Conclusions

This chapter provided a comprehensive literature review on the ever-evolving subject of needs of twenty-first-century engineers. The review spans over significant number of studies worldwide, and no relevant investigation has been detected in the Middle East or the GCC area. Synthesis of the literature led to development of a four-dimensional model of engineering skills that consist totally of 22 competencies. The model has been utilized in subsequent stages of the investigation in order to solicit perspectives of stakeholders in any country of investigation, for current and emergent needs of engineering talent attributes. Furthermore, other systematic models representing global twenty-first-century engineering talent skills and competency development models have been developed and described in detail.

Synthesis of the literature on needs of engineering education system for twenty-first-century engineering graduates has been provided. The next chapter provides further details on leadership and engineering.

References

2028 Vision for Mechanical Engineering: A Report of the Global Summit on the Future of Mechanical Engineering. (2008). *Notes*. New York, USA. Retrieved from www.asme.org.

Abdullah, S., Zaharim, A., Harris, S. M., Omar, M. Z., Basri, H., & Mohamed, N. A. N. I. K. (2007). Engineering education: Using technical attributes to analyse the employers' expectation of future engineering graduates in Malaysia. *Engineering Education, 39*, 191–195.

ACER. (2002). *Employability skills for Australian industry: Literature review and framework development*.

Agogino, A. M. (2005). *Educating the engineer of 2020: Global visions for the new century NAE engineer of 2020.*

Allan, M., & Chisholm, C. U. (2009). The formation of the engineer for the 21 st century—A global perspective. In *20th Australasian Association for Engineering Education Conference* (pp. 447–452).

Anderson, D., Somerville, M., Berbeco, H., Bourne, J. R., Crisman, J., Dabby, D., et al. (2005). The Olin curriculum: Thinking toward the future. *IEEE Transactions on Education, 48*(1), 198–205.

ASCE. (2007). *The vision for civil engineering in 2025*. Virginia. Retrieved from www.asce.org.

Barrie, S. C. (2007). A conceptual framework for the teaching and learning of generic graduate attributes. *Studies in Higher Education, 32*(4), 439–458.

Bourn, D., & Neal, I. (2008). *The global engineer: Incorporating global skills within UK higher education of engineers. Engineer.*

Bowman, K. (2010). Background paper for the AQF council on generic skills (March).

Calfee, C. (2009). Galloway' s 21st century engineer: An essay review Robert Calfee University of California, Riverside Stanford University Thomas Stahovich University of California, Riverside. In P. D. Galloway (2008) *The 21st Century Engineer: A Proposal for E, 12*(14).

Canada, C. B. (2000). *Employability skills 2000+*. Canada.

Casner-Lotto, J., & Barrington, L. (2006). *Are they really ready to work? Employers' perspectives on the basic knowledge and applied skills of new entrants to the 21st century US workforce.* Retrieved from http://www.eric.ed.gov/ERICWebPortal/recordDetail?accno=ED519465.

CBI. (2007). *Time well spent Embedding employability in work experience*. London, UK.

CBI. (2009). *Jobs for the future: The business vision for sustainable employment in the UK*. UK.

CBI. (2012). *Learning to grow: What employers need from education and skills—Education and skills survey 2012*. London, UK. Retrieved from www.cbi.org.uk.

Chan, A. D. C., & Fishbein, J. (2009). Aglobal engineer for the global community. *The Journal of Policy Engagement, 1*(2), 4–9.

Chang, Y., Atkinson, D., & Hirleman, E. D. (2009). International research and engineering education: Impacts and best practices education: Impact and best practices. *Online Journal for Global Engineering Education, 4*(2), 1.

Childs, P., & Gibson, P. (2010). Graduating professional engineers and management skills-are they adequate for the workplace? *3rd International Symposium for Engineering Education, 2010, University College Cork, Ireland*, 1–9. Retrieved from http://ro.uow.edu.au/chsd/34/.

Continental. (2006). *Final report of the global engineering excellence initiative; educating the next generation of engineers for the global workplace* (p. 114). Hanover/Germany.

Danielson, S. (2011). ASME vision 2030: Helping to inform mechanical engineering education. In *41st ASEE/IEEE frontiers in education conference, 12–15, 2011* (pp. 1–6). Rapid City, SD. Retrieved from http://ieeexplore.ieee.org/xpls/abs_all.jsp?arnumber=6143065.

Dawe, S. (2002). *Focussing on generic skills in training packages*. Australia. Retrieved from http://citeseerx.ist.psu.edu/viewdoc/download?doi=10.1.1.198.3625&rep=rep1&type=pdf.

DEST. (2006). *Employability skills from framework to practice: An introductory guide for trainers and assessors*. Melbourne.

DEST. (2007). *Graduate employability skills: Prepared for the business, industry and higher education collaboration council*. Melbourne.

Duderstadt, J. J. (2008a). *Engineering for a changing world a roadmap to the future of a roadmap to the future of engineering*.

Duderstadt, J. J. (2010). Holistic engineering education. In: D. Grasso & M. B. Burkins (Eds.) (pp. 17–35). doi:10.1007/978-1-4419-1393-7.

Duderstadt, J. J. (2011). *Heartland papers a master plan for higher education in the Midwest*. Chicago, Illinois, USA.

Dudman, A., & Wearne, S. (2003). Professional engineers' needs for managerial skills and expertise. *Centre for Research in the Management of Projects University of Manchester Institute of Science & Technology*, (March). Retrieved from http://www.engc.org.uk/ecukdocuments/internet/documentlibrary/ProfessionalEngineersneedManagementSkills.pdf.

Edinburgh. (2012). *Edinburgh Napier University Conscious Employability Model*. Edinburgh. Retrieved from http://www.napier.ac.uk/oldcontent/standoutfromthecrowd/Pages/ConsciousEmployabilityModel.aspx.

ERT. (1995). *Education for Europeans—Towards the learning society*. Brussels: The European Round Table of Industrialists.

Felder, R. M. (1998). ABET criteria 2000: An exercise in engineering problem solving. *Chemical Engineering Education, 32*(2), 126–127.

Finegold, D., & Notabartolo, A. (2010). 21st-century competencies and their impact: An interdisciplinary literature review. *Research on 21st Century Competencies, National Research Council*, 1–50. Retrieved from http://onlinelibrary.wiley.com/doi/10.1002/cbdv.200490137/abstract.

Fung, M. C. D., Lee, W. Y. W., & Wong, S. L. P. (2006). A new measure of generic competencies.

Gibb, J. (2004). *Generic skills in vocational education and training: Research readings*. National Centre for Vocational Education Research. Retrieved from http://www.ncver.edu.au.

Hager, P., & Holland, S., (2006), *Graduate attributes, learning and employability* (pp. 1–15). Springer. http://www.springer.com/education+%26+language/book/978-1-4020-5341-2.

Holsti, O. R. (1969). *Content analysis for the social sciences and humanities*. Reading, MA: Addison-Wesley.

Hounsell, D. (2011). *Graduates for the 21st Century: Integrating the enhancement themes—Institutional activities*. Edinburgh. Retrieved from www.enhancementthemes.ac.uk.

HRA. (2013). *It Takes More Than a Major: Employer Priorities for College Learning and Student Success an Online Survey Among Employers Conducted on Behalf of: The Association of American Colleges And Universities* (pp. 1–15). Washington, DC, USA.

Hundley, S. (2012). *AC 2012-4233: Attributes of a global engineer: Field-informed perspectives, recommendations, and implications*. Retrieved from http://www.asee.org/public/conferences/8/papers/4233/view.

IEA. (2009). *Graduate attributes and professional competency* (pp. 1–15). Retrieved from http://www.washingtonaccord.org/GradProfiles.cfm.

IFEES. (n.d.). The attributes of a global engineer project. In *International Federation of Attributes of a Global Engineer Project Focus Group Event at WEEF, Engineering Education Societies* (pp. 1–10). Buenos Aires, Argentina.

Jamieson, L., Brophy, S., House, N., Harris, M., Delaurentis, D., Howell, K., et al. (2009). *Purdue's Engineer of 2020: The Journey, in 2009 Annual Conference & Exposition; New Learning Paradigms II*.

Kastenberg, W., Hauser-Kastenberg, G., & Norris, D. (2006). An approach to undergraduate engineering education for the 21st century. In *Proceedings of the Frontiers in Education 36th Annual Conference* (pp. 23–28). doi:10.1109/FIE.2006.322502.

Kerr, I. R. (2010). Futures thinking for engineering and engineers Australia's continuing professional development process. *Australasian Journal of Engineering Education, Institution of Engineers Australia, 16*(1), 13–20.

Khair, R., Abdulwahed, M., Hamouda, A., and Hasna, M.O., (2013). From Global to Local: *Investigation of Necessary Engineering Skills for KBE Transformation in Qatar in the Context of Global Engineering Attributes* (pp. 637–643). In Frontiers in Education Conference, October, Oklahoma, USA, IEEE.

King, R. (2008). *Engineers for the future: Addressing the supply and quality of Australian engineering graduates for the 21st century* (pp. 1–137). ISBN: 978-0-9805211-0-8.

Kirkpatrick, A. T. (2011). AC 2011-2804: Vision 2030 creating the future of mechanical engineering education.

Klein-gardner, S. S. (2011). Defining global competence for engineering students. In *118th ASEE 2011 Annual Conference & Exposition*. BC, Canada.

Knight, D. B. (2012). In search of the engineers of 2020: An outcome-based typology of engineering undergraduates, AC 2012-3337. In *Proceedings of the 119th Annual Conference of the American Society for Engineering Education*, San Antonio, TX.

Kolbe, R. H., & Burnett, M. S. (1991). Content-analysis of an for examination improving applications with directives and reliability objectivity. *The Journal of Consumer Research, 18* (2), 243–250.

Li, T., & Cavusgil, S. (1995). A classification and assessment of research streams in international marketing. *International Business Review, 4*(3), 251–277.

Llanes, C. C. (2008). *Outcomes-based education: The TIP experience* (pp. 1–7). Technological Institute of the Philippines.

Male, S. A. (2010a). Perceptions of competency deficiencies in engineering graduates. *Australasian Journal of Engineering Education, 16*(1), 55–68. Retrieved from http://www.engineersmedia.com.au/journals/aaee/pdf/AJEE_16_1_MaleF2.pdf.

Male, S., & Chapman, E. (2005). Assessing the generic competencies of engineering graduates: Preliminary report from an ongoing research program. In *Proceedings of the 2005 ASEE/AaeE 4th Global Colloquium on Engineering Education.* Australasian Association for Engineering Education. Retrieved from http://www.oii.uwa.edu.au/__data/assets/pdf_file/0012/1898229/Assessing-the-generic-competencies-of-engineering-graduates.pdf.

Markes, I. (2006). A review of literature on employability skill needs in engineering. *European Journal of Engineering Education, 31*(6), 637–650. doi:10.1080/03043790600911704.

Martin, R., Villeneuve-smith, F., Marshall, L., & Mckenzie, E. (2008). *Employability skills explored. ...: Learning and skills .... London*. Retrieved from http://norfolkunites.org.uk/media/pdf/ese.pdf.

Mena, I. B., Zappe, S. E., & Litzinger, T. A. (2012). *Preparing the Engineer of 2020: Analysis of Alumni Data, in 2012 ASEE Annual Conference*. American Society for Engineering Education.

Michigan. (2009). *Michigan Engineering 2020, The Commission on Undergraduate Engineering Education: Curriculum for the 21st Century. Rivers.* Retrieved from http://www.engin.umich.edu/admin/adue/undergradcommission/.

Mishra, S. (2010). Engineering curricula in the 21st century: The global scenario and challenges for India. *Journal of Engineering, Science and Management Education, 1*(1), 29–33.

MSA. (2012). *Foundation skills in the manufacturing industry* (pp. 1–20).

NAE. (2004). *The engineer of 2020: Visions of engineering in the new century.*

NAE. (2005). Educating the engineer of 2020: Adapting engineering education to the new century. *IEEE Engineering Management Review*,. doi:10.1109/EMR.2009.4804343..

NAE. (2011). *Assessing 21st century skills*. Washington, DC.

NCVER. (2003). *Defining generic skills at a glance* (pp. 1–12).

NESP. (2010). *Evaluation report: National employability skills program*. Canada. Retrieved from http://scholar.google.com/scholar?hl=en&btnG=Search&q=intitle:Evaluation+Report:+National+Employability+Skills+Program#1.
Nguyen, D. (1998). The essential skills and attributes of an engineer: A comparative study of academics, industry personnel and engineering students. *Global Journal of Engineering Education*, *2*(1), 65–76. Retrieved from http://citeseerx.ist.psu.edu/viewdoc/download?doi=10.1.1.124.1502&rep=rep1&type=pdf.
Nguyen, N. D., Yoshinari, Y., & Shigeji, M. (2005). University education and employment in Japan: Students' perceptions on employment attributes and implications for university education. *Quality Assurance in Education, 13*(3), 202–218. doi:10.1108/09684880510607945.
NHEAP. (2007). *Malaysia national higher education action plan 2007–2010*.
Noakes, N. (2004). *Employability, lifelong learning, personal development planning and eportfolios at HKUST* (pp. 1–8). Retrieved from http://repository.ust.hk/dspace/handle/1783.1/1710.
Nor, N. M., Rajab, N., & Ismail, K. M. (2008). Educating the engineer of 2020: Malaysian scenario. In *International Conference on Engineering Education: New Challenges in Engineering Education and Research in the 21st Century*. Budapest, Hungary.
NRC. (2007). Rising Above The Gathering Storm: Energizing and Employing America for a Brighter Economic Future Committee on Prospering in the Global Economy of the 21st Century: An Agenda for American Science and Technology. National Academy Press, Washington, DC.
NSB. (2007). *Moving forward to improve engineering education* (Vol. NSB-07–122, p. 58). National Science Board. ISBN: NSB-07-122.
Obama's campaign to improve the participation and performance of America's students in science, technology, engineering, and mathematics (STEM). "Educate to Innovate". (2009). [Online]. http://www.whitehouse.gov/issues/education/educate-innovate. Accessed June 26, 2011.
OECD. (2002). *Definition and selection of competencies (DeSeCo) theoretical and conceptual foundations: Strategy paper*. Organisation for Economic Co-operation and Development.
P21. (2008). *21st century skills, education & competitiveness: A resource and policy guide*. Tucson, USA. Retrieved from http://www.pkwy.k12.mo.us/projectParkway/File/21st_century_skills_education_and_competitiveness_guide.pdf.
Palmer, B., Terenzini, P. T., McKenna, A. F., Harper, B. J., & Merson, D. (2011). Design in context: Where do the engineers of 2020 learn this skill. In *Proceedings of the 118th Annual Conference of the American Society for Engineering Education* (Vol. 2020). Vancouver, Canada. Retrieved from http://146.186.208.241:8082/Plone/e2020/asee-conference-presentations/asee2011-contextual-competence-final-paper.pdf.
Parkinson, A. (2009). The rationale for developing global competenc. *Online Journal for Global Engineering Education*, *4*(2). Retrieved from http://digitalcommons.uri.edu/ojgee/vol4/iss2/2.
Patil, A., & Codner, G. (2008). Global accreditation for the global engineering attributes: A way forward. In *Proceedings of the 2008 AaeE Conference, Yeppoon* (pp. 1–8).
Pinelli, T. E., & Haynie, W. J. (2010). A case for the nationwide inclusion of engineering in the K-12 curriculum via technology education. *Journal of Technology Education, 21*(2), 52–68.
PyeTait. (2011). *Ready for work? The work readiness of entrants to the UK Construction sector who have work-related qualifications* (pp. 1–72). Harrogate, UK. Retrieved from www.pyetait.com.
Rabl, M., & Hillmer, G. (2012). The cultivation of engineering talent. In *SEFI 40TH annual conference*. Thessaloniki, Greece. Retrieved from http://www.sefi.be/conference-2012/Papers/Abstracts/040sum.pdf
RAE. (2007). *Educating engineers for the 21st century educating engineers for the 21st century*. London: The Royal Academy of Engineering.
Rajala, S. A. (2012). Beyond 2020: Preparing engineers for the future. In *Proceedings of the IEEE, 100* (Special Centennial Issue) (pp. 1376–1383). doi:10.1109/JPROC.2012.2190169.
Townsend, R., & Waterhouse, P. (2008). *Whose responsibility? Literature Review & Methodology —Support document*.

REFLEX. (2007). *The flexible professional in the knowledge society: General results of the REFLEX project.* Netherlands. Retrieved from http://ideas.repec.org/p/ner/maastr/urnnbnnlui27-20754.html.
Saravanan, V. (2006). Sustainable employability skills for engineering professionals. *The Indian Review of World, 5*(2), 1–9. Retrieved from http://www.worldlitonline.net/sustainable-employability-skills-for-engineering-professionals.pdf.
SCANS. (1991). *What Work Requires of Schools: A SCANS Report for America 2000. The Secretary's Commission on Achieving Necessary.* Washington, DC. Retrieved from http://scholar.google.com/scholar?hl=en&btnG=Search&q=intitle:WHAT+WORK+REQUIRES+OF+SCHOOLS+:+A+SCANS+REPORT+FOR+AMERICA+2000#0.
Shah, A. A. (2009). *Contributions and limitations of self assessment of competences by higher education graduates.* Universite de Bourgogne.
Sheppard, K., Dominick, P., & Aronson, Z. (2004). Preparing engineering students for the new business paradigm of international teamwork and global orientation. *International Journal of Engineering Education, 20*(3), 475–483. Retrieved from http://citeseerx.ist.psu.edu/viewdoc/download?doi=10.1.1.162.3040&rep=rep1&type=pdf.
Shuman, L. J., Atman, C. J., Eschenbach, E. A., Evans, D., Felder, R. M., Imbrie, P. K., et al. (2000a). The future of engineering education. I. A vision for a new century. *32nd Annual Frontiers in Education, 34*(1), T4A-1–T4A-15. doi:10.1109/FIE.2002.1157986.
Shuman, L. J., Atman, C. J., Eschenbach, E. A., Evans, D., Felder, R. M., Imbrie, P. K., et al. (2000b). The future of engineering education III. Developing critical skills. *32nd Annual Frontiers in Education, 34*(2), T4A–1–T4A–15. doi:10.1109/FIE.2002.1157986.
Shuman, J., Besterfield-Sacre, M., & McGourty, J. (2005). The ABET 'professional skills'—Can they be taught? Can they be assessed? *Journal of Engineering Education, 49*(1), 41–55.
Skills, W., & They, W. (2009). *Transportation engineers of 2020: What skills will they need?* (pp. 1–8).
Spinks, N., Silburn, N., & Birchall, D. (2006). *Educating engineers for the 21st century: The industry view. Engineering.* Greenlands, UK.
Statz, C. (2003). *Two perspectives for assessing skills: What can they contribute to policy.* Paper prepared for EGFSN, Sept. 2003.
Sunthonkanokpong, W. (2011). Future global visions of engineering education. *Procedia Engineering, 8*, 160–164. doi:10.1016/j.proeng.2011.03.029.
Swearengen, J., & Barnes, S. (2002). Globalization and the undergraduate manufacturing engineering curriculum. *Journal of Engineering Education, 49*(1), 255–261. doi:10.1002/j.2168-9830.2002.tb00700.x/abstract.
Tech, V. (2011). The emergence of engineering education research as an internationally connected field of inquiry. *Journal of Engineering Education, 100*(1), 14–47.
Terenzini, P. T., & Lattuca, L. R. (2011) Benchmarking U.S. engineering education Vis-à-vis the engineer of 2020. In *Convocation of the Professional Engineering Societies.*
Tippins, N., & Hilton, M. (2010). *A database for a changing economy: Review of the occupational information network (O* NET).* Washington, DC. Retrieved from http://books.google.com/books?hl=en&lr=&id=Qw1CxQ_QiV4C&oi=fnd&pg=PT5&dq=A+Database+for+a+Changing+Economy+:+Review+of+the+Occupational+Information+Network+(+O+*+NET+)&ots=Y7DnLmrMnu&sig=zCWUW91ZCn8qNneU551DIFFYmVg.
Tryggvason, G., & Apelian, D. (2006). Re-engineering engineering education for the challenges of the 21st century. *IEEE Engineering Management Review, 58*, 14–17.
UKCES. (2009). *The employability challenge full report.* London. Retrieved from ukces.org.uk.
UNESCO. (2012). *Graduate employability in asia* (pp. 1–96). Bangkok: UNESCO.
Walther, J., & Radcliffe, D. F. (2007). The competence dilemma in engineering education: Moving beyond simple graduate attribute mapping. *Australasian Journal of Engineering Education, 13* (1), 41–51.
WDA. (2009). The Singapore employability skills system (August).

Wearne, H. (1984). *Managerial skills and expertise used by samples of engineers in Britain.* Western Canada, Australia, Japan, the Netherlands and Norway. Report TMR 152, Technological Management, University of Bradford.

Wheelahan, L., Moodie, G., & Briggs, C. (2011). *Rethinking skills in vocational education and training: From competencies to capabilities*. Melbourne, Australia. Retrieved from http://www.voced.edu.au/td/nsw_68.10.

Yorke, M., & Knight, P. T. (2006). *Employability embedding employability into the curriculum.* UK. ISBN: 1-905788-00-2.

Zaharim, A., Yusoff, Y. M., Mohamed, A., Omar, M. Z., Muhamad, N., & Mustapha, R. (2010). Practical framework of employability skills for engineering graduate in Malaysia. In *Education Engineering Conference (EDUCON), 14–16 April 2010* (pp. 921–927). Madrid: IEEE. doi:10.1109/EDUCON.2010.5492478.

Zaharim, A., Yusoff, Y., & Omar, M. (2009). Engineering employability skills required by employers in Asia. In *Proceedings of the 6th WSEAS International Conference on Engineering Education* (pp. 195–201). Retrieved from http://www.wseas.us/e-library/conferences/2009/rodos/EDU/EDU30.pdf.

Chapter 3
Leadership: Models, Competencies, and the Emergence of Engineering Leadership

3.1 Introduction to Leadership

There are many definitions of leadership, and each definition has a different area of focus (Nauman and Khan 2006). One of the ways to describe leadership is the "skills approach", which places emphasis on the skills or abilities that can be learned or developed (Northouse 2010). Leadership is sometimes defined as the ability to inspire confidence and support among people who are needed to achieve organizational goals (Kan Kim and Maubourgne 1992). Thus, leadership is about determining direction and influencing people about values, vision, mission, and strategy (Reeve 2010). "Other definitions are "interpersonal influence", directed through communication towards goal attainment", "the art of influencing people by persuasion or example to follow a line of action", and "the principle dynamic force that motivates and coordinates the organization in the accomplishment of its objectives" (Dubrin 2007).

Leadership skills are in high demand, such skills are sought in candidates for high-level and entry-level professional positions, and they were listed as the top attributes executives seeking in candidates after industry-specific experience and functional/technical expertise. Komives and her colleagues argue that leadership, like any other skill, needs to be learned and practiced (Komives et al. 1998). Astin (1993) argues that it is important to develop young men and women during their college years to become future leaders.

3.2 Effective Leaders

Leaders are people who act as guides for groups of people bringing them together to unleash their potential. They are risk-taking and able to learn from mistakes and failures (Bonasso 2001). The best leaders are inspiring visionaries, with flair and

M. Abdulwahed and M.O. Hasna, *Engineering and Technology Talent for Innovation and Knowledge-Based Economies*,
DOI 10.1007/978-3-319-46439-8_3

attractive personalities. Effective leaders also possess efficient management skills as essential for enabling realization of the visions and missions they drive. Effective leaders empathize with people, and always aim to take advantage of new opportunities. Effective leaders are masters of emotional intelligence, and they apply it in order to get the best out of each member of the team (Chieh 2012).

3.3 Characteristics and Attributes of Leadership

Characteristics associated with leadership can be classified into three categories: personality traits, motives, and cognitive factors. These required characteristics for leadership are caused by a combination of heredity and environment (Dubrin 2007).

3.3.1 Personality and Interpersonal Traits

Observations by managers and human resource specialists, as well as many research studies, indicate that leaders have certain personal and interpersonal traits. (Kirkpatrick and Locke 1991; Goleman 1998). In general, personality traits of effective leaders may encompass the following attributes: 1—self-confidence, 2—humility, 3—assertiveness, 4—trustworthiness, 5—warmth, 6—sense of humour, 7—enthusiasm, 8—extraversion, 9—emotional stability, 10—management, 11—leadership, 12—communications, 13—teamwork, etc.

3.3.2 Motives

Effective leaders have frequently been distinguished by their motives and needs. Leaders are known for working hard and putting forth energy to achieve their goals, this is called the drive and achievement motives (Dubrin 2007). Leaders with high power of motives have three dominant characteristics: 1—they act with vigour and determination to exert their power, 2—they invest much time in thinking about ways to alter the behaviour and thinking of others, and 3—they care about their personal standing with those around them (McClelland and Byatzis 1982).

3.3.3 Cognitive Factors

Mental ability as well as personality is important for leadership success. To inspire people, bring about constructive change, and solve problems creatively, leaders need to be mentally sharp. Competencies such as **problem-solving** skills and

intellectual skills are referred to collectively as cognitive factors (Dubrin 2007). A meta-analysis of 151 studies found a positive relationship between **intelligence** and job performance of leaders in different settings (Judge et al. 2004). On the other hand, intellectual ability is closely related to having knowledge of the business or the key task the group is performing. Leaders at every level are expected to bring forth useful ideas for carrying out the mission of the organization. Furthermore, such knowledge is particularly important when developing strategy and formulating mission statements.

Creativity is also such an important aspect of the leader's role in the modern organization. Many effective leaders are creative in the sense that they arrive at imaginative and original solutions to complex problems.

Another important cognitive trait of leaders is **insight**, a depth of understanding that requires considerable intuition and common sense. Intuition is often the mental process used to provide the understanding of a problem. Insight into people and situations helps leaders make the best use of both their own and other's talents and speed decision-making for wise choices.

To develop visions and corporate strategy, a leader needs **far-sightedness**: the ability to understand the long-range implications of actions and policies, and **conceptual thinking**: the ability to see the overall perspective and make far-sightedness possible. A conceptual thinker is also a systems thinker because he/she understands how the external environment influences the organization and how different parts of the organization influence each other.

Yet another important cognitive characteristic of leaders is their **openness to experience**, or their positive orientation towards learning. Leaders who have a great deal of openness to experience have well-developed intellects including imagination, curiosity, and intelligent artistic sensation (Dubrin 2007).

3.4 Leadership Models

There are many models of leadership that were investigated and argued in many studies such as transformational leadership, servant leadership, ethical leadership, and authentic leadership. In this section, we will be discussing a few examples of these models.

3.4.1 Transformational Leadership

Transformational leaders are those who are able to inspire followers to change expectations, perceptions, and motivations to work towards common goals. A transformational leadership style is mainly based on leader's personality traits and ability to energize challenging goals for followers, believing that they can do their best. Transformational leaders have 4 main personality traits, referred to as the 4 I's transformational leadership model (Bass and Riggio 2008).

- **Idealized Influence (II)**—the leader serves as an ideal role model for followers,
- **Inspirational Motivation (IM)**—transformational leaders have the ability to inspire and motivate followers. Combined, these first two I's are what constitute the transformational leader's charisma.
- **Individualized Consideration (IC)**—Transformational leaders demonstrate genuine concern for the needs and feelings of the followers. This personal attention to each follower is a key element in bringing out their very best efforts.
- **Intellectual Stimulation (IS)**—the leader challenges followers to be innovative and creative.

3.4.2 Servant Leadership

Servant leadership is defined/characterized by placing the needs of followers before the personal interests of the leader (Farling et al. 1999; Patterson 2003; Russell and Stone 2002). Servant leadership is increasingly popular in leadership studies and practices. Servant leadership is argued to hold potential in changing organizations and societies as it stimulates both personal and organizational metamorphoses (Russell and Stone 2002). It is pointed to the servant leader as a person who is sharply different from one who is leader first. "Becoming servant leaders engages us in personal, internal self-change, and changes outward behaviour" (Matteson and Irving 2006). Patterson (2003) presented servant leadership theory as an extension of transformational leadership theory. Patterson's (2003) model of servant leadership includes the following dimensions as essential characteristics of servant leadership: love, humility, altruism, vision, trust, empowerment, and service.

3.4.3 Ethical Leadership

Brown et al. (2005) have defined ethical leadership as "the demonstration of normatively appropriate conduct through personal actions and interpersonal relationships, and the promotion of such conduct to followers through two way communication, reinforcement, and decision making" (Aronson 2001; Kanungo 2001; Trevino et al. 2003). When the leader's moral integrity is in doubt, the leader will more likely fail to influence followers to achieve organizational goals (Kanungo 2001).

Drawing on this definition of ethical leadership and in line with previous research (Trevino et al. 2003; Brown et al. 2005) describes ethical leaders as **honest**, **trustworthy**, **fair,** and **caring**. Such leaders make principled and fair choices and structure work environments justly. Other components are a leader's

transparency, **engagement in open communication** with followers, and clarification of expectations and responsibilities so that employees are clear on what is expected from them (Kanungo 2001).

3.4.4 Other Models

There are several other leadership models in the literature, such as charismatic leadership, laissez-fair leadership, task-related leadership, relationship-oriented leadership, participative leadership, autocratic leadership, entrepreneurial leadership, situational leadership, normative decision leadership, and cognitive resource theory leadership.

3.5 Emergence of Leadership in Engineering

Today's fast-paced global economy, with the scope, difficulty, and complexity of problems, mandates a critical need for leaders engineers (Jablokow 2008). Every engineering organization, big or small, needs leaders and managers for the success of the organization and projects they execute. (Kumar and Hsiao 2007). Engineers are hired (among other functions) to solve complex problems, perform detailed analysis of processes/systems and structures, to design, and to develop and implement (Reeve 2010). Engineering leadership can be defined as the ability to lead a group of personnel from various profiles for creating, designing, developing, implementing, and evaluating products, systems, or services (Crumpton-Young et al. 2010). Engineering leadership is also defined as management of technical change, innovative conception, design, and implementation supported by the invention of enabling technologies to meet the needs of customers and society. (Northouse 2010).

Several researchers argue that leadership is an important skill that should be included in engineering education curricula (Cox et al. 2009; Kumar and Hsiao 2007). The importance of leadership education/training in engineering has been highlighted in various initiatives and reports, including those by, the National Academy of Engineering (NAE 2004, 2005), and the works of Komives et al. (2007), Kouzes and Posner (2008), Kotnour et al. (2014).

According to the NAE (2004), "engineers must understand the principles of leadership and be able to practice them in growing proportions as their careers advance". Engineers need to exercise their leadership due to "the growing interdependence between technology and the economic and social foundations of modern society". Criteria for accrediting engineering programs for USA in 2012–2013 include not only strong analytical skills, but also skills indirectly related to leadership (e.g. the ability to communicate effectively, to function in multidisciplinary teams, and to understand the impact of engineering solutions in global and societal contexts). These and other reports (Hilton 1995; McCuen 1999) have called for better preparation of engineering undergraduates to meet the leadership demands of a changing global society (Ahn et al. 2014). Experts believe that every young engineer must be given the opportunity to lead in their early career and must be allowed to manage a project and be encouraged to develop soft skills, such as interpersonal, marketing, and communications skills (Bowman and Farr 2000). In order to be a true engineering leader, engineering students must possess technical and nontechnical soft skills, which would give them an edge in the workplace (Kumar and Hsiao 2007). They must possess skills such as written and oral communication, customer relations, personal initiative, teamwork abilities, organizational knowledge, and decision-making that will facilitate the development of solutions to business challenges, to be effective leaders (Newport and Elms 1997).

A critical re-examination of the current engineering education system suggests a possible reformation of the current system. One of the most essential leadership attributes is teamwork; yet, studies show that students are lacking these skills (Adams 2003; Bahner 1996; Gardner 2001; Natishan et al. 2000). Many researchers found that many undergraduate students within engineering departments do not receive leadership or management skills training that are needed for them to succeed as leaders within future engineering positions (Kumar and Hsiao 2007).

Promoted in the form of minors, formal undergraduate degree programs, formal graduate degree programs, and graduate courses, leadership has been identified as a skill that needs to be included in the curricula for future engineers (Cox et al. 2009). Russell and Yao agreed that "an engineer is hired for her or his technical skills, fired for poor people skills, and promoted for leadership and management skills" (Russell and Yao 1996).

3.6 Engineering Leadership' Education, Programs, and Training

Leadership development should be as an integral part of the educational program of engineering colleges worldwide. To meet this current need, this can best be accomplished by various strategies as promising steps forward in that context. Many engineering leadership education and development programs have been utilized in different universities and countries. These programs are typically composed of coursework, workshops, seminars, and assignments.

Examples of international good practice in engineering leadership education are grouped into several categories; **explicit programs**, where engineering leadership development is the primary explicit objective; **non-explicit programs**, where the engineering leadership development is embedded within a broader remit (Graham et al. 2009); and **corporate** programs.

3.6.1 Explicit Academic Programs

One of the good practices with an "**explicit**" leadership objective is the *Gordon-MIT Engineering Leadership Program at MIT.* The program combines a suite of educational activities to develop the leadership capabilities of the undergraduate engineers at MIT with a broader national role to improve the leadership capabilities of future US engineering graduates. In pursuit of these goals, the program has already secured external funding of $20 m. Another example is the engineering leadership program at Iowa State University and Leadership in a Technological Environment at Monash University, Australia. This program offers a three-year co-curricular leadership development experience to the top tier of engineering students with the highest entry qualifications.

3.6.2 Non-explicit Academic Programs (Embedded)

On the other hand, one example of good practice with a "non-explicit" leadership agenda includes the *EWB Challenge* coordinated by *Engineers Without Borders Australia* where first-year engineering students from universities across Australia are provided with a real design brief from the developing world and asked to produce "engineering and design solutions that are tailored to the local social, cultural, political, environmental, and economic context".

3.6.3 Corporate Programs on Leadership

There are several engineering leadership development programs available for current professional engineers offered by various industry leaders such as Lockheed Martin, National Instruments, Raytheon, and GE (Lockheed 2005; National Instruments 2007; Raytheon 2006; GE 2009). These programs offer leadership development tracks to qualified new engineering employees, focused on facilitating the transition from an academic to a corporate mindset and are composed of leadership training, rotational assignments, and career development. For instance, the Lockheed Martin program consists of lectures, team projects, and leadership development conferences designed to provide engineers with practical and strategic leadership and interpersonal skills ("Lockheed Martin Leadership Development Program" 2005). GE offers several engineering leadership programs that focus on the company's lifelong learning goal for its employees (GE 2009).

Researchers highlighted that engineering leadership programs should facilitate developing attributes, such as the ability to manage a group, have critical thinking, and be visionary, inspirational, influential, adaptable, open-minded, people-centred, action-oriented, equitable, interpersonal, determined, confident, a good communicator, accountable, fair, and a networker; to ensure that engineering professionals and future engineers are prepared to be competent leaders (Crumpton-Young et al. 2010).

3.7 Design Projects and Leadership

3.7.1 Design Projects as a Vehicle for Leadership Training

Team design projects can effectively help in developing engineering students' leadership skills (Chieh 2012). Besides, projects help students sharpen their interpersonal skills, and industry experiences help by providing the practical knowledge and mentoring that engineering students do not receive in the classroom (Crumpton-Young et al. 2010). Huemann et al. (2007) suggest that the project is a social system and includes several areas focused on organizational behaviour, leadership, communication, team building, and human resource management. A number of European engineering schools have developed peer tutoring models with a strong leadership element, where more senior students "coach" junior student project teams. An example of such a model is the ***Project Management in Practice*** (Ozgen et al. 2008) ***course at the Universitat Rovirai Virgili, Spain***, where selected fourth year students lead first-year group design project teams (Graham, nd.).

Schools are increasingly relying on group design projects wherein students learn to apply leadership skills, such as team building and communicating. Providing students with "real" projects for "real" sponsors greatly enriches traditional education by bringing students' textbooks to "life". Projects afford undergraduates a transition from textbooks to the workplace wherein they learn to apply their new-found knowledge (Bowman and Farr 2000).

3.7.2 Leadership Attributes for Design Projects Success

Several leadership-related attributes that have been correlated with design projects success are as follows: effective leadership (Ammeter and Dukerich 2002), good communication (Pettersen 1991; White and Fortune 2002), and the ability to operate under pressure in a complex environment (Pettersen 1991; White and Fortune 2002). Verma (1995) lists the following people skills that are important for project managers, apart from the technical knowledge and decision-making skills that they require: communication, motivation and negotiation, self-confidence, reliability, maturity and emotional stability, a constructive, positive attitude, and flexibility and tolerance for ambiguity and uncertainty. Kerzner (2003) states that effective leaders are not completely task or relationship focused in their actions, rather they maintain a balance between the two.

3.8 Conclusions

The chapter provided a panoramic overview of leadership attributes, models, and recent emergences in engineering leadership. Several institutions have been progressively embracing engineering leadership development whether implicitly or through explicit programs. Specific details on the relationship between team-based complex engineering design and leadership development have been provided.

References

Adams, S. G. (2003). Building successful student teams in the engineering classroom. *Journal of STEM Education, 4*(3/4), 1–6.

Ahn, B., Cox, M. F., London, J., Cekic, O., & Zhu, J. (2014). Creating an instrument to measure leadership, change, and synthesis in engineering undergraduates. *Journal of Engineering Education, 103*(1), 115–136.

Ammeter, A. P., & Dukerich, J. M. (2002). Leadership, team building, and team member characteristics in high performance project teams. *Engineering Management Journal, 14*(4), 3–8.

Aronson, E. (2001). Integrating leadership styles and ethical perspectives. *Canadian Journal of Administrative Sciences, 18*, 244–256.

Astin, A. (1993). *What matters in college?* San Francisco: Jossey-Bass.

Bahner, B. (1996). Report: Curricula need product realization. *ASME News, 15*(10), 1–6.

Bass, B. M., & Riggio, R. E. (2008). *Transformational leadership*. New Gersy: The Taylor and Francis e-library.

Bonasso, S. G. (2001). Engineering, leadership, and integral philosophy. *Journal of Professional Issues in Engineering Education and Practice, 127*, 17–25.

Brown, M. E., Trevino, L. K., & Harrison, D. A. (2005). Ethical leadership: A social learning perspective for construct development and testing. *Organizational Behavior and Human Decision Processes, 97*, 117–134.

Bowman, B. A., & Farr, J. V. (2000). Embedding leadership in civil engineering education. *Journal of Professional Issues in Engineering Education and Practice, 126*, 16–20.

Chieh, H. C. (2012). *Leadership skills for Engineers*. National University of Singapore (NUP).

Cox, M. F., Berry, C. A., & Smith, K. A. (2009). Development of a leadership, policy, and change course for science, technology, engineering, and mathematics graduate students. *Journal of STEM Education, 10*(3–4), 9–16.

Crumpton-Young, L., McCauley-Bush, P., Rabelo, L., Meza, K., & Ferreras, A. (2010). Engineering leadership development programs a look at what is needed and what is being done. *Journal of STEM Education, 11*(3/4), 10.

Dubrin, J. (2007). *Leadership: Research findings, practice, and skills* (5th ed.). Boston, New York: Houghton Mifflin Co.

Farling, M. L., Stone, A. G., & Winston, B. E. (1999). Servant leadership: Setting the stage for empirical research. *The Journal of Leadership Studies, 6*, 49–72.

Gardner, P. D. (2001). Job opportunities for the next 30 years. *The Black Collegian* (April), 54–58.

General Electric (GE) Leadership Development Programs. (2009). http://www.gepower.com/about/careers/en/leadership_programs/index.htm.

Graham, R., Crawley, E., & Mendelsohn, B. R. (2009). Engineering leadership education: A snapshot review of international good practice. *MIT Engineering Leadership Program.*

Goleman, D. (1998). What makes a leader? *Harvard Business Review, 82*, 92–102.

Hilton, M. (1995). The importance of civil engineering leadership in the government sector. *Journal of Management in Engineering, 11*(5), 12–16.

Huemann, M., Keegan, A. E., & Turner, J. R. (2007). Human resource management in the project oriented company: A critical review. *International Journal of Project Management, 25*, 312–320.

Jablokow, K. W. (2008). Developing problem solving leadership: A cognitive approach. *International Journal of Engineering Education, 24*(5), 936–954.

Judge, T. A., Colbert, A. E., & Ilies, R. (2004). Intelligence and leadership: A quantitative review and test of theoretical propositions. *Journal of Applied Psychology, 89*, 542–552.

Kan Kim, W., & Maubourgne, R. A. (1992) Parables of leadership. *Harvard Business Review* 123.

Kanungo, R. N. (2001). Ethical values of transactional and transformational leaders. *Canadian Journal of Administrative Sciences, 18*, 257–265.

Kerzner, H. (2003). *Project management: A systems approach to planning, scheduling and controlling*. New York: John Wiley and Sons.

Kirkpatrick, S. A., & Locke, E. A. (1991). Leadership: Do traits matter? *Academy of Management Executive, 5*(2), 48–60.

Komives, S., Lucas, N., & McMahon, T. (1998). *Exploring leadership: For college students who want to make a difference*. San Francisco: Jossey-Bass.

Komives, S. R., Lucas, N., & McMahon, T. R. (2007). *Exploring leadership for college students who want to make a difference* (2nd ed.). San Francisco, CA: Jossey-Bass.

Kouzes, J. M., & Posner, B. Z. (2008). *The student leadership challenge: Five practices for exemplary leaders*. San Francisco, CA: Jossey-Bass.

Kotnour, T., Hoekstra, R., Reilly, C., Knight, R., & Selter, J. (2014). Infusing leadership education in the undergraduate engineering experience: A framework from Ucf's Eli2. *Journal of Leadership Studies, 7*, 48–57.

Kumar, S., & Hsiao, K. (2007). Engineers learn "soft skills the hard way": Planting a seed of leadership in engineering classes. *Leadership and Management in Engineering, 7*(1), 18–23.

Lockheed Martin Leadership Development Program. (2005). http://www.lockheedmartin.com/wms/findPage.do?dsp=fec&ci=12964&sc=400.

Matteson, J., & Irving, J. (2006). Servant versus self-sacrificial leadership: A behavioural comparison of two follow-oriented leadership theories. *International Journal of Leadership Studies, 2*(1), 36–51.

McClelland, D. C., & Byatzis, R. (1982). Leadership motive pattern and long-term success in management. *Journal of Applied Psychology, 67*, 727.

McCuen, R. H. (1999). A course on engineering leadership. *Journal of Professional Issues in Engineering Education and Practice, 125*(3), 79–82.

NAE. (2005). *Educating the engineer of 2020: Adapting engineering education to the new century*. Washington, DC: The National Academies Press.

NAE. (2004). *The engineer of 2020: Visions of engineering in the new century*. Washington, DC: The National Academies Press.

National Instruments Engineering Leadership Program. (2007). http://digital.ni.com/jobs.nsf/$$opportunitiesalpha/C61A6084C59DFCDB862572730057EA74?OpenDocument.

Natishan, M. E., Schmidt, L. C., & Mead, P. (2000). Student focus group results on student team performance issues. *Journal of Engineering Education, 89*(3), 269–272.

Nauman, S., & Khan, A. M. (2006). Patterns of leadership for effective project management. *Journal of Quality and Technology Management 20*.

Newport, C. L., & Elms, D. G. (1997). Effective engineers. *International Journal of Engineering Education, 13*(5), 325–332. http://www.ijee.dit.ie/articles/Vol13-5/ijee996.pdf.

Northouse, P. G. (2010). *Leadership: Theory and practice* (5th ed.). Thousand Oaks, CA: Sage.

Ozgen, S., Alabart, J. R., & Medir, M. (2008). A 360°-degree feedback process to assist senior engineering students in their leadership development. In *SEFI 36th Annual Conference on Quality Assessment, Employability and Innovation, July 2–5*, Aalborg, Denmark.

Patterson, K. (2003). Servant leadership: A theoretical model. *Dissertation Abstracts International, 64*(02), 570.

Pettersen, N. (1991). What do we know about the effective project manager? *International Journal of Project Management, 9*(2), 99–104.

Raytheon's Engineering Leadership Development Program (LDP) Targets HighPotential Employees. (2006). http://www.raytheon.com/feature/new06_ldp.

Reeve, D. W. (2010). There is an urgent need for engineering leadership education. *Engineering Leadership Review 1*(1), 1–6.

Russell, R., & Stone, A. G. (2002). A review of servant leadership attributes: Developing a practical model. *Leadership & Organization Development Journal, 23*(3), 145–157.

Russell, J. S., & Yao, J. T. P. (1996). Education conference delivers initiatives. *Journal of Management in Engineering ASCE, 12*(6), 17–26.

U.S. News and World Report On Line. (1998). http://www.usnews.com/USnews.

Trevino, L. K., Brown, M., & Hartman, L. P. (2003). A qualitative investigation of perceived executive ethical leadership: Perceptions from inside and outside the executive suite. *Human Relations, 56*, 5–37.

Verma, V. K. (1995). *Organising projects for success: The human aspect of project management* (Vol. 1). Upper Darby: PMI.

White, D., & Fortune, J. (2002). Current practice in project management-an empirical study. *International Journal of Project Management, 20*(1), 1–11.

Chapter 4
On Industry—Academia Collaboration for iKBE

4.1 Introduction

It is noticed that an evolution of perception and implementation of academia–industry linkage is emerging from operational activities towards more of a strategic partnership (in many cases facilitated by governmental programmes) for producing joint knowledge and economical outcomes. The engineering, science, and technology domains play the vast majority of roles in the emerging roles of twenty-first-century universities and their engagement and transfer relationships with industry and society (Davey et al. 2011).

Several terminologies are used in the literature to describe the interactions, transactions, and relations between the academic and the industrial/business sectors, such as "The Education—Business Cooperation," "Industry–Academia Linkage," "Industry–University Cooperation," "Technology Transfer," "Knowledge Transfer," and "Innovation Diffusion,". We coin the term technology, innovation, and knowledge and assets transfer (TIKAT) as a holistic representation for academia socio-economic engagement and impact for industry/business, government, and society. In this chapter, we provide details of some well-developed frameworks from Europe and the USA; these frameworks are briefly described in the next sections.

4.2 The Triple-Helix Model of Stanford

The Triple-Helix framework was developed in mid-1990s and started to have recognition from early 2000 in Stanford in order to conceptualize the emerging roles of twenty-first-century universities beyond teaching and research into entrepreneurial and socio-economic impact driver. The model puts significant emphasis on the integrative role of universities together with government and industry in

M. Abdulwahed and M.O. Hasna, *Engineering and Technology Talent for Innovation and Knowledge-Based Economies*,
DOI 10.1007/978-3-319-46439-8_4

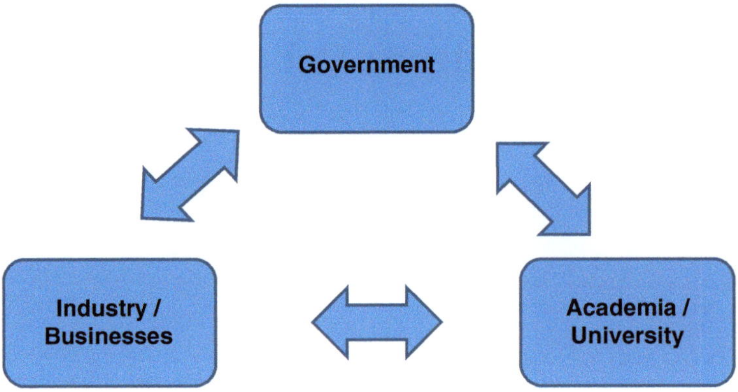

Fig. 4.1 The Triple-Helix model from Stanford; *source* taken from (Davey et al. 2011)

order to share future socio-economic growth. Figure 4.1 shows a representation of the Triple-Helix model.

4.3 The Industry—University Cooperative Research Centre (I/UCRC)

The I/UCRC framework and model was originated three decades ago from an NSF programme initiated to foster academia–industry collaboration in research and development. The programme has been highly successful and is running to present day. The NSF (2016) describes the programme as: "I/UCRC enables industrially-relevant, pre-competitive research via a multi-member, sustained partnerships between industry, academe, and government. NSF supports the development and evolution of I/UCRCs, providing a financial and procedural framework for membership and operations in addition to best practices learned over decades of fostering public/private partnerships that provide significant value to the nation, industry and university faculty and students. I/UCRCs bring together their members (faculty, students, industry stakeholders) to perform cutting-edge pre-competitive fundamental research in science, engineering, technology area(s) of interest to industry and that can drive innovation and the U.S. economy. Members guide the direction of Center research through active involvement and mentoring."

Throughout the decades of experience, NSF developed best practices and design models in establishing I/UCRC, and these are well detailed in Gray and Walters (1998, edited). NSF provides fine details, instruments, and tools related to I/UCRC, such as planning and initiating new centre, foundational principles for effective organizational structure, membership composition, planning and implementing the

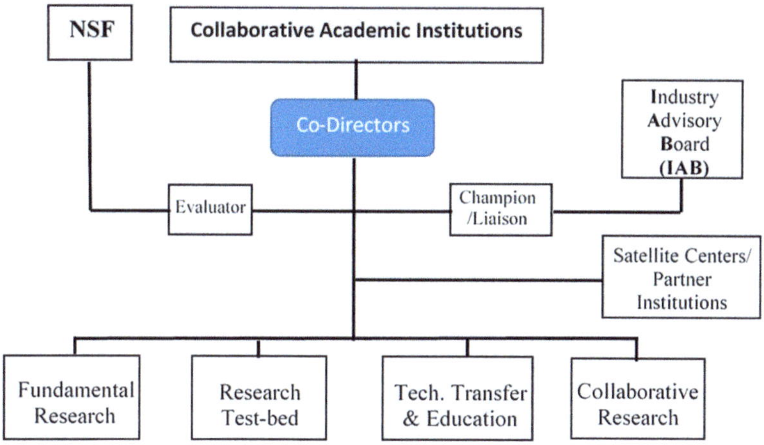

Fig. 4.2 Organizational and operational structure of the intelligent maintenance center—an industry–university cooperative research centre; *source* taken from (IMS 2015)

collaborative research programmes, communications and marketing, control and budgeting, knowledge and technology transfer, and finally centre-effective leadership. Figure 4.2 shows an example of one of the I/UCRC centres "Intelligent Maintenance Systems".

4.4 Framework 3: The University—Business Cooperation (UBC)

The UBC was a result of the largest scale study on collaboration between academia and sectors of government, industry and employers, and societal system enterprises (referred to collectively as "*Business*"). The study covered all countries of the European Economic Area (EEA); this included 33 countries and 6000+ samples using a mixed methods approach associated with secondary reviews and comprehensive cases analysis from across Europe. UBC was defined as "all types of direct and indirect, personal and non-personal interactions between Higher Education Institutions (HEIs), and business for reciprocal and mutual benefits" (Davey et al. 2011). UBC results/outcomes were defined beyond the creation of patents, licence, and research contracts. In total, in eight major areas the UBC was defined to occur, and the areas are as follows:

- R&D collaboration
- Mobility of academics
- Mobility of students
- Commercialization of R&D results

- Curriculum development and delivery
- Lifelong learning
- Entrepreneurship
- Management.

The study resulted in a systemic strategic and implementation framework for enhancing UBC, and the framework is adopted later on in this document for CENG with slight modifications in terminologies for focusing on engineering and technology. The term "Business" in the UBC framework refers to all potential interactions and transactions (mutual way) between HEIs and industry, government, and social enterprises/stakeholders (Davey et al. 2011); the majority of UBC identified in the study took place within the engineering, science, and technology domains.

The UBC has its grass roots in the Triple-Helix model from Stanford (originated early in 2000) that describes the triple interaction between academia–industry/business-government as the foundational driver of socio-economic development.

4.4.1 Main Findings of the UBC

Several findings were identified, such as the following:

1. UBC is crucial for creating a knowledge society; see Fig. 4.3 for direct and indirect benefits leading to a iKBE/KBS
2. There are eight types of UBC that are more direct, measurable, and having promotable benefits (e.g. collaboration in R&D and mobility of students in internships);
3. Situational factors such as age, previous experience with industry, and gender have contributions to UBC;
4. Lack of funding and excess of bureaucracy at all levels are the highest barriers to UBC;
5. Personal relationships drive UBC, and it is people-driven;
6. The creation and development of supporting mechanisms is critical for UBC.

4.4.2 The UBC Ecosystem

The UBC ecosystem is composed of 6 main strategic components (key stakeholders, supporting mechanisms, influencing factors, university–business cooperation, outcomes for HEIs/academics/industry, and finally knowledge society), see Fig. 4.3. Effective implementation of the previous levels will lead ***to***

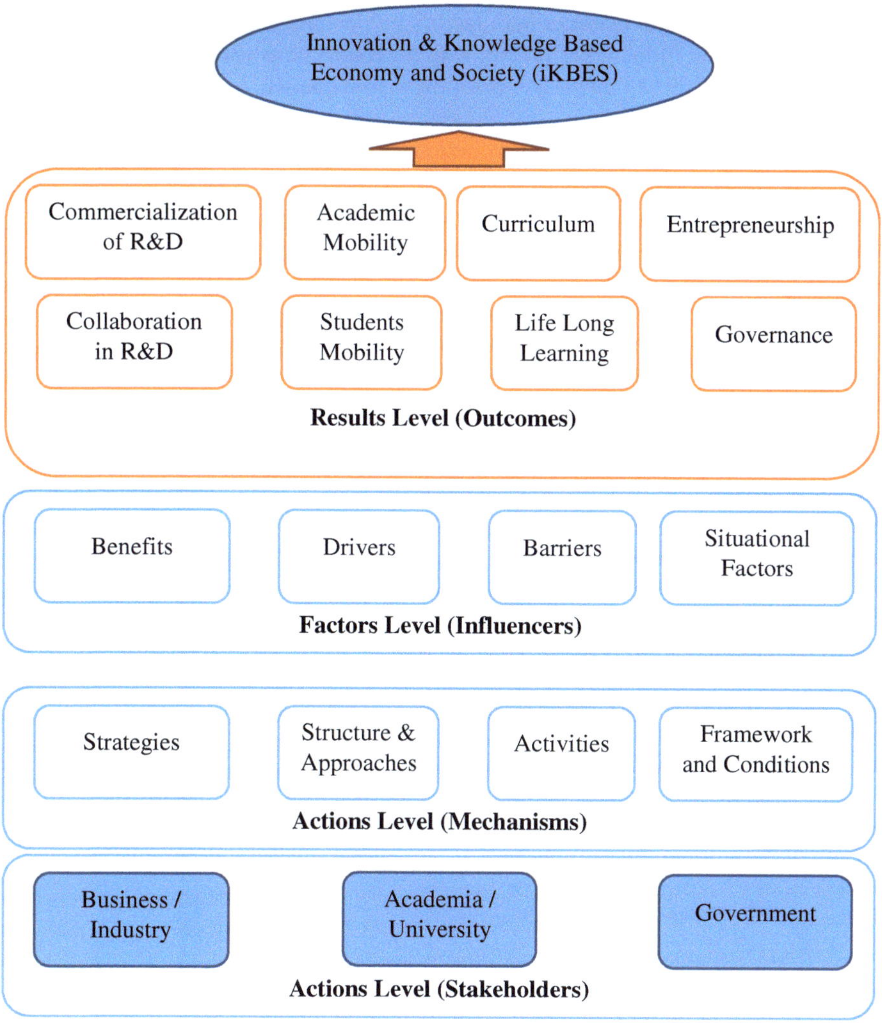

Fig. 4.3 Strategic components of UBC ecosystem and levels/dimensions; *source* taken from Davey (2011)

socio-economic development, or what is called a Knowledge-Based Society and Economy. The UBC ecosystem is composed of several levels (Davey 2011):

- **Result and Outcome Level/Dimension**: where the extent of UBC is visible, this level contains 8 main different types of UBC that can be undertaken (1—Collaboration in R&D, 2—mobility of academics, 3—mobility of students, 4—commercialization of R&D results, 5—curriculum development and delivery, 6—lifelong learning, 7—entrepreneurship, and 8—governance)

- **Factor Levels/Dimension**: where factors influencing UBC efforts are located, it contains all the influencing factors which need to be considered in any attempt to affect UBC; these factors are as follows: 1—benefits, 2—drivers, 3—barriers, and 4—situational factors
- **Action Level/Dimension**: where actions to stimulate UBC occurs, this area contains four pillars where actions to influence the extent of UBC occur (1—strategies, 2—organizational structure and approaches, 3—operational activities, and 4—framework conditions).

4.5 The TIKAT Framework, an Adoption of UBC for Engineering Schools and Colleges

The UBC can be adopted for engineering colleges and schools. For the context of engineering adoption of UBC, we will re-label it as a technology innovation and knowledge and assets transfer (TIKAT) framework to match more specifically with engineering and technology terms of industry–academia collaboration.

The "Stairway model for strategic partnership" (see Fig. 4.4) is one of the systemic ways of implementing gradual transfer from an initial phase of formation, through basic collaboration, joint collaboration, and finally a strategic partnership. The model has been utilized successfully by several institutions in Germany and the UK (Davey et al. 2011); below is a brief description of the model phases:

1. **Formation**: In the formation phase, the regularity of collaboration is limited to only initiation of contact, and having some common understandings of mutual benefits maybe also with some formalities put in place in terms of an MoU.

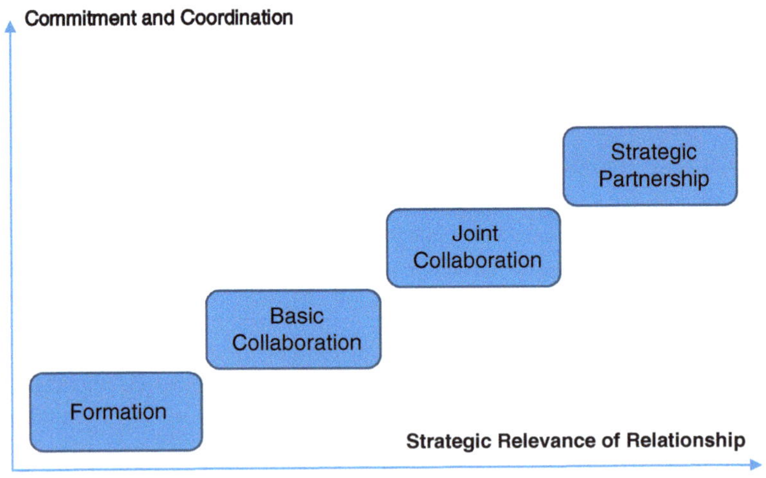

Fig. 4.4 The "Stairway model for strategic partnership" is a model composed of four stages, leading to a strategic collaboration; *source* taken from Davey et al. (2011)

Normally, activities initiated in this phase are of a short-term nature, and potentially as pilots. Few people are usually engaged in such contacts and done on the bottom line of the organizational hierarchy of academic institutions by a professor or even a teaching assistant.
2. **Basic Collaboration**: In this stage, one or more of concrete projects are launched and running; however, no regular pattern of such projects is established yet. Time horizon is short to medium and also few people were involved, mainly done on a faculty/professor level.
3. **Joint Collaboration**: In this stage also, one or more concrete projects are established; however, a regular pattern of collaboration started to emerge. Time horizon of this phase is medium to long as both sides are experiencing mutual added value of the joint collaboration. Normally, senior management on departmental or college level (e.g. head of department and/or dean) are involved in managing the joint work with industry.
4. **Strategic Partnership**: Here, it is similar to the previous phase with the exception that relationship has moved into a sustainable long-term partnership and is normally managed on a university senior leadership level.

4.5.1 Pillar 1—Strategies: Top Strategies for the Implementation UBC/TIKAT Ecosystem in Engineering Schools

Davey et al. (2011) identified the following top impactful strategies for supporting the implementation of UBC/TIKAT ecosystem (arranged by the order of impact and slightly modified to fit engineering colleges context):

1. **Pillar 1. Strategy 1**: The inclusion of TIKAT (UBC) as part of the assessment of engineering academic work and performance;
2. **Pillar 1. Strategy 2**: The dedication of resources (including funding) to support UBC/TIKAT;
3. **Pillar 1. Strategy 3**: The provision of incentives for academics to encourage UBC/TIKAT;
4. **Pillar 1. Strategy 4**: The internal promotion of UBC/TIKAT in the college;
5. **Pillar 1. Strategy 5**: The external promotion of UBC/TIKAT with external partners;
6. **Pillar 1. Strategy 6**: A documented vision/mission embracing UBC/TIKAT in the college new strategy;
7. **Pillar 1. Strategy 7**: Building a holistic strategy for UBC/TIKAT in the college strategy.

4.5.2 Pillar 2—Structure and Approaches: Top Structures and Approaches for the Implementation of UBC/TIKAT Ecosystem in Engineering Schools

Davey et al. (2011) identified the following top impactful structures and approaches for supporting the implementation of UBC ecosystem (arranged by order of impact):

1. **Pillar 2. Structure and Approach 1**: Industry professional employed in knowledge transfer area in the college;
2. **Pillar 2. Structure and Approach 2**: Technology ventures hub for development of new tech. business innovations;
3. **Pillar 2. Structure and Approach 3**: College board member for UBC/TIKAT;
4. **Pillar 2. Structure and Approach 4**: Engineering alumni network;
5. **Pillar 2. Structure and Approach 5**: Existence of career office;
6. **Pillar 2. Structure and Approach 6**: Existence of engineering academic on business board of industry partners;
7. **Pillar 2. Structure and Approach 7**: Business people on college board;
8. **Pillar 2. Structure and Approach 8**: External agencies dedicated to UBC/TIKAT;
9. **Pillar 2. Structure and Approach 9**: Internal agencies dedicated to UBC/TIKAT;
10. **Pillar 2. Structure and Approach 10**: Existence of a dedicated contact person, contact agency, or a programme/initiative for mobility of academics, mobility of students, lifelong learning, and/or Technology Entrepreneurship.

4.5.3 Pillar 3—Operational Activities: Top Operational Activities for the Implementation of the UBC Ecosystem

Davey et al. (2011) identified the following top impactful operational activities for supporting the implementation of UBC ecosystem (arranged by the order of impact):

1. **Pillar 3. Operational Activity 1**: Collaboration activities facilitating student interaction with industry;
2. **Pillar 3. Operational Activity 2**: Workshops, information sessions, and forums for UBC/TIKAT targeting academics;
3. **Pillar 3. Operational Activity 3**: Technology entrepreneurship education offered for students;

4. **Pillar 3. Operational Activity 4**: Collaboration activities facilitating academics interaction with industry;
5. **Pillar 3. Operational Activity 5**: Networking sessions and frequency of meetings between academics and industry stakeholders;
6. **Pillar 3. Operational Activity 6**: Technology entrepreneurship training offered for academics.

4.5.4 Pillar 4—Framework: Top Framework Elements for the Implementation of the UBC Ecosystem

Davey et al. (2011) identified the following top impactful operational activities for supporting implementation of UBC ecosystem (arranged by the order of impact):

1. **Pillar 4. Framework 1**: Laws/regulations positively supporting UBC (TIKAT);
2. **Pillar 4. Framework 2**: Laws/regulations allowing movement of staff between the college and industry;
3. **Pillar 4. Framework 3**: Laws/regulations positively supporting the creation of new technology-based companies and ventures.

4.6 Industry Advisory Boards (IAB) for Engineering Schools: An Advanced Industry Advisory and Strategic Partnership (IASP) Model

Industry advisory boards are becoming increasingly part of engineering education schools; the movement was in particular pushed through the ABET accreditation agency that required engineering colleges and departments to have an industry advisory board. In this section, an advanced model of an industry advisory board that aims at developing strategic partnership is provided. In this model, the IAB is recalled as Industry Advisory and Strategic Partnership (IASP) board.

4.6.1 IASP Board Vision

The IASP board vision is to formulate a competent engineering and technology partnership model with industry and government as a catalyst for sustainable socio-economic development.

4.6.2 IASP Board Mission

The IASP board mission is to transform linkage and collaboration, between college of engineering and industry–government sector, into a strategic partnership at all related aspects of ***mutual benefits and national, regional, or global priority***.

4.6.3 IASP Board Strategic Objectives

There are three main strategic objectives of the IASP board:

1. **Strategic Objective 1**: Partner for the development of engineering and technology capacity and talent;
2. **Strategic Objective 2**: Foster mutual transfer of technology, innovation, knowledge, experience, and assets between college of engineering and industry–government sector;
3. **Strategic Objective 3**: Fund-raising resources for the development of partnership implementation strategic initiatives and projects.

4.6.4 IASP Board Strategic Partnership Areas of Focus

There are four main strategic partnership areas of focus in which implementation actions, strategic initiatives, and projects will be geared into:

1. **Area 1—Capacity Building**: partnership for capacity building and talent management:
   (a) Engineering/STEM education and outreach in K-12;
   (b) Engineering and technology education in the college;
   (c) Engineering and technology professional and lifelong learning programmes;
2. **Area 2—Research and Development**: Partnership in joint applied research and decolopment:
   (a) Engineering and technology policy-level research (education, innovation, linkage, talent management, and development);
   (b) Applied technical development for industry problems;
   (c) Joint research and development centres and innovation hubs (iHubs);
3. **Area 3—Transfer**: mutual technology, innovation, knowledge, and assets transfer (m-TIKAT):
   (a) Mobility/transfer of academics, industry professionals, and students;

(b) Maximization of utilization of resources and assets in all sectors (expertise, laboratories, equipment, workshops, and seminars);
(c) Technology innovation and entrepreneurship (in large enterprise entrepreneurial activities), and entrepreneurship (start-ups and SMEs focus).

4.6.5 IASP Board Structure and Process of Development

The IASP board structure is composed of the following:

1. **IASP Steering Board**: Strategic committee composed of top executives and decision-makers;
2. **IASP Implementation Board**: Implementation committee composed of senior managers and technical expertise;
3. **IASP Funds**: Funds for sponsoring agreed on joint strategic initiatives and projects.

Process management and implementation of the IASP board is composed of the following subsystems:

- IASP strategic guidelines and decision-making system;
- IASP planning and implementation system
- IASP quality assurance and evaluation system.

4.6.6 IASP Funds

The IASP funds can be established as a finance contribution from industry and government organizations for funding strategic initiatives/projects emerging from the IASP board decisions in order to implement its vision, mission, and strategic objectives. The IASP funds will be managed by the IASP steering board jointly with the college of engineering.

4.6.7 IASP Steering Board Membership

The IASP board is composed of senior-level and decision-making executives from the following:

- **Government**
- **Industry**

- **Health Sector**
- **Finance Sector**
- **Mega Projects and Initiatives (National, regional, or global)**
- **Private Sector**
- **Professional Organizations**
- **College of Engineering Management**
- **Third Party**:
  - Regional and international expertise from academia;
  - Regional and international expertise from industry;
  - Regional and international expertise from government.

4.6.8 IASP Implementation Board

The supplementary board is the implementation management arm of the IASP board. Each of the IASP board members will elect a representative for the implementation board. Actions and strategic projects/initiatives agreed on by the IASP board will be communicated to the implementation board for execution and management of implementation. The implementation board will meet twice a semester, and implementation subgroups may meet more when needed.

4.7 Conclusions

Collaboration and integration of universities with industry and government is a core concept in iKBEs. The transformation into innovation- and knowledge-based economies implies universities transfer their role from only teaching and research into more entrepreneurial institutions with socio-economic impact dimensions.

Several frameworks and models have emerged in the last two decades in order to foster university collaboration and partnership with industry; in particular, this chapter provided details on the Industry-Advisory Board model, Triple-Helix model, the Industry–University Cooperative Research Center model, and the university–business cooperation framework. Furthermore, this chapter provided adoptions and extensions tailored in particular for engineering schools and colleges.

References

Davey, T., Baaken, T., Muros, V. G., & Meerman, A. (2011). *The state of European University-business cooperation: Final report–study on the cooperation between higher education institutions and public and private organisations in Europe*. Science-to-Business Marketing Research Centre Germany: Muenster.

Gray, D. O., & Walters, S. G. (1998). *Managing the industry/university cooperative research center*. Columbus, OH: Battelle.

NSF. (2016). NSF—National Science Foundation. http://www.nsf.gov/.

Chapter 5
On the Response for iKBE Development: Excellence in Technology' Innovation, Entrepreneurship, & Engineering/Education (TIEE), and a Proposed Structure

5.1 Innovation, Design, and Technology Entrepreneurship for iKBE

5.1.1 Global Trends

Innovation, design, and commercialization/entrepreneurship have recently been fostered in novel engineering curricular and education models as vehicles to integrate foundational, scientific, technical, and socio-economic aspects associated with engineering and engineering learning, producing in a meaningful product-/service-oriented manner. Innovation and design have been introduced in first-year engineering programmes and in multidisciplinary activities/courses/degrees in novel engineering curricular and education models; see, for instance, models at Purdue University. Singapore is the only innovation and knowledge-based economy (iKBE) that has programmed clear innovation and design schemes and is known for its higher education technical institutions focusing on innovation and design; recently, a state-of-the-art Singapore University of Technology and Design (SUTD) was established in partnership with MIT. The new university is the latest investment of the Singapore government that focuses on developing high-quality engineering education driven by multidisciplinary technological innovation and design. The Singapore government's philosophy of focus on technology innovation and design has led to higher value outcomes and GDP growth compared to similar iKBE countries.

It has also led to higher growth in engineering publications despite relatively lower expenditure on R&D than similarly developed countries (Luo et al. 2011). Figure 5.1 shows Singapore's growth in R&D expenditure as a percentage of GDP over 20 years; it is significantly lesser than other countries such as Finland or South Korea. Figure 5.2 shows GPS growth in Singapore, which is significantly higher than other comparative countries.

M. Abdulwahed and M.O. Hasna, *Engineering and Technology Talent for Innovation and Knowledge-Based Economies*,
DOI 10.1007/978-3-319-46439-8_5

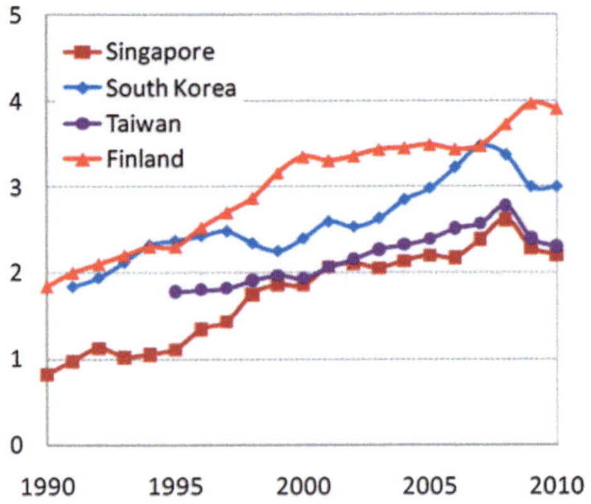

Fig. 5.1 Singapore R&D expenditure as percentage of GDP (Luo et al. 2011)

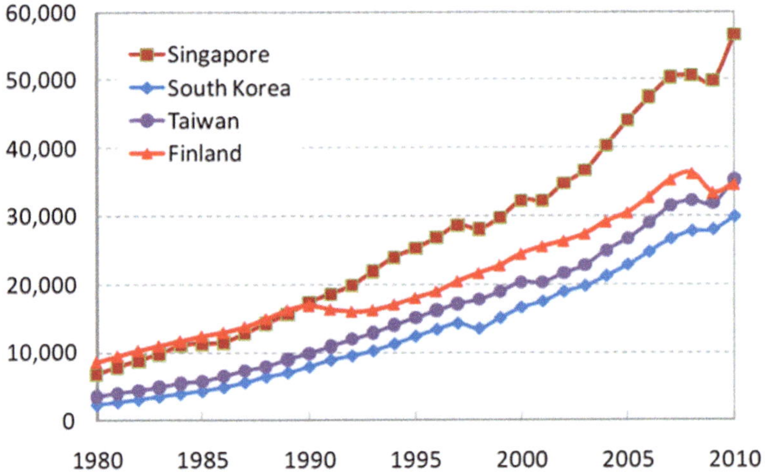

Fig. 5.2 Singapore GDP per capital (Luo et al. 2011)

5.1.2 Local Perspectives

Over the past 2 years, the interest of engineering students in local, regional, and international extra-curricular activities related to tech. innovation and entrepreneurship has significantly increased. Some of existing engineering students already started up some companies.

Increasingly, an innovation, design, and entrepreneurship-oriented engineering education curriculum is becoming more and more adopted in iKBE-based engineering education institutions in various advanced countries in the world. The pressing needs in Qatar may dedicate the urgent needs for systematic integration of these aspects into the structural and curricular means of existing engineering education institutions in Qatar.

5.1.3 Design as a Vehicle for Twenty-First-Century Competencies Development

While the major purpose of integrating engineering design in engineering education is to develop solid technical understanding and skills, engineering design may have other significant contributions to twenty-first-century soft skills development (Abdulwahed et al. 2014). A large-scale study on complex engineering design competition across Europe, the Middle East, and Asia (A; Abdulwahed et al. 2016) has demonstrated statistically significant higher level of satisfaction of the vast majority of twenty-first-century engineering competencies via complex engineering design projects as compared with conventional engineering education curriculum in colleges of engineering. The findings were universal in Europe, Middle East, and Asia, and also consistent with genders (Appendix J; Table A.21 through Table A.24 illustrated as Figs. 5.3 and 5.4). Basic engineering design and thinking has been utilized as an innovative and twenty-first-century competencies development enabler in several non-engineering arenas. Most notably, in business education, there is increasingly more training included on design thinking. In the USA, formal engineering design education and practice in fore and more deployed in K-12 system. Regardless whether K-12 students will continue in engineering studies or careers later, engineering design in K-12 aimed also to develop twenty-first-century competencies of future generations.

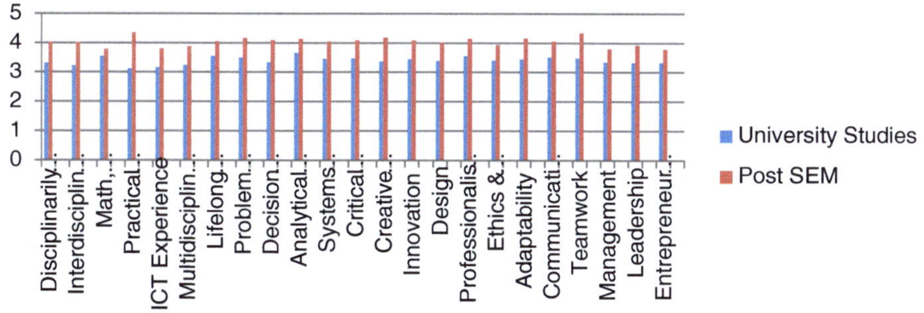

Fig. 5.3 Twenty-first-century engineering competencies complex engineering design (SEM) project as compared to conventional engineering education curriculum in colleges of engineering "Genders perceptions"

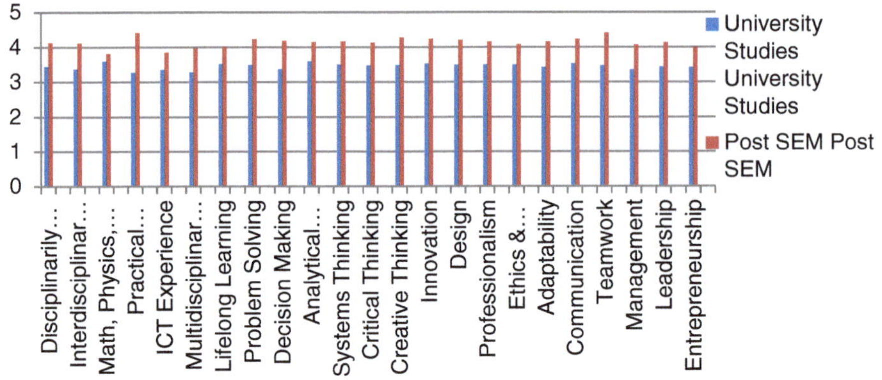

Fig. 5.4 Twenty-first-century engineering competencies complex engineering design (SEM) project as compared to conventional engineering education curriculum in colleges of engineering "Origin perception"

5.2 The Technology, Innovation, and Entrepreneurship Driving Force Triangle

We call the triplet composed of government, industry, and academia as the technology, innovation, and entrepreneurship triangle; this is in particular based on the Triple-Helix model but focused further on the tech. entrepreneurship context. The higher integration, communications, and coordination among the three sectors, the faster pace and better outcomes of establishing a technology and engineering innovation system, SMEs, and processes for a innovation and knowledge-based economy. Each of these sectors possesses a number of enabling and complementary assets for a well-rounded tech. innovation and entrepreneurial ecosystem, see Fig. 5.5. Close collaboration and integration among the three sectors for creating rapidly grown and effective innovation and entrepreneurial ecosystem is highly important. The next subsection provides some details on related assets in each sector.

5.2.1 Government

Governments set up plans and visions transform into a innovation and knowledge-based economy and also implement strategic development plans in order to meet this objective. Furthermore, enabling legislations are progressively being approved in the process of enhancing the entrepreneurial ecosystem iKBEs (e.g. the case of Finland and Singapore). The governments overall own or manage policies, programmes, incentives, and finance necessary for implementing and enhancing a technological innovation and entrepreneurial ecosystem in the country.

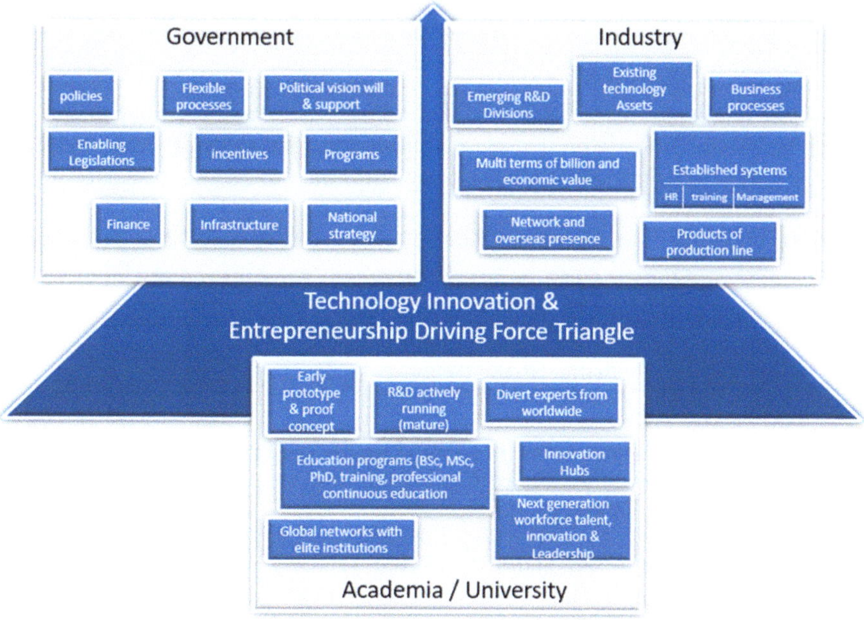

Fig. 5.5 Technology, innovation, and entrepreneurship triangle

5.2.2 *Industry*

Industries are the engine of iKBE transformation. For the case of Finland, industry was transformed from machinery and manufacturing into highly innovative ICT and services, and in the case of Singapore, industry was transformed from labour-intensive into technology-intensive, and lately into ICT, biomedical, and services.

Qatar industry is to a large extent dominated by large enterprises (in particular, in the oil and gas sectors), which is counter to the case of innovation and knowledge-based economies where small and medium-sized enterprises (SMEs) count for around 85–90 % of a country's GDP. In our perspectives, Qatar's large enterprises could play a crucial and driving role in widening the scale of SMEs in Qatar through leveraging their own capabilities in spinning off new start-ups. Qatar's large enterprises possess a number of established assets that can significantly accelerate the implementations of new start-ups, and such assets include (but are not limited to) the following: emerging innovation and R&D divisions, existing technology and information systems, mature business processes, significant liquidity in terms of finance and funds, established HR/training/management systems, production lines, networks, and overseas presence.

5.2.3 Academia

Academia is the base member of the technology innovation and entrepreneurship ecosystem. Innovation-based technical and engineering academic entities can provide an incubation of early prototypes and proof concepts. Overall, R&D is actively running in engineering academic institutions, attributed mainly to the establishment of research funds and the various grant programmes they manage every year. Academia is normally well networked locally and internationally. Academia overall develops and manages educational programmes (B.Sc., M.Sc., and Ph.D.) as well as training programmes, and graduates next generation of technology and engineering talent.

5.3 Incepting Disruptive Transformations in a College of Engineering for a iKBE: The TIEE Concept

The Technology Innovation and Engineering Education (TIEE) is an evolving organizational engineering and technology structure in Qatar University, see concepts' details in (Abdulwahed and Hasna 2012; Abdulwahed et al. 2015). The TIEE concept (interchangeably utilized in this book is also as an acronym of technology innovation, entrepreneurship, and education) has been developed in particular taking Qatar National Vision 2030 and the transformation into a iKBE at the very core, and to enable the college to better respond to these emerging needs of the country. In this section, the concept of TIEE is provisioned as an example of organizational structure development for other engineering education institutions that aim to play key roles in innovation and development. The concept is derived from similar global examples; it is generic and can be deployed elsewhere with slight modifications. The TIEE concept is in particular vital for engineering institutions in the Middle East as a vehicle for economic and technological development.

5.3.1 International Benchmarking

The last 10–15 years have witnessed escalation in the development of departments and centres focusing on multidisciplinary technology innovation, design, and engineering education, mainly in the USA but noticeably worldwide. These entities are dedicated entirely to academic provisioning, practice, and research in topics such as innovation, design, and engineering education. Examples of their academic offerings are as follows: first-year engineering programmes, engineering design experience, multidisciplinary engineering minors/majors or degrees, and structured

M.Sc. and Ph.D. programmes in related areas; see Table 5.1 for examples of activities and programmes of TIEE similar entities.

The most notable examples are the School of Technology, Design, and Professional Engineering Education at Pennsylvania State University, and the

Table 5.1 Sample of multidisciplinary engineering organizational entities

Entity and housing college	Name	Scope	Brief description	Examples of departments in the same institution
School of Engineering Education College of Engineering—Purdue University	Pre-College Engineering	K-12	Outreach, research, and projects related to engineering in K-12	School of Civil Engineering School of Chemical Engineering School of Industrial Engineering School of Mechanical Engineering School of Electrical and Computer Engineering
	First-Year Engineering Programme	Freshman	First-Year Engineering Programme	
	Idea to Innovation (i2i)	UG	Innovation and Design Integrative Space and Mentoring	
	Multidisciplinary Engineering Programme with Concentration in General Engineering	UG	B.Sc. Degree (ABET Accredited) with concentration on General Engineering	
	Multidisciplinary Engineering Programme with Concentration in Engineering Management	UG	B.Sc. Degree (ABET Accredited) with concentration on Engineering Management	
	Engineering Education	PG	Ph.D. Degree	
School of Engineering Design, Technology, and Professional Programmes College of Engineering—Pennsylvania State University	First-Year Engagement	Freshman	Freshman courses for First-Year Engineering Experience Engagement	Department of Architectural Engineering Department of Chemical Engineering Department of Computer Science and Engineering Department of Industrial and Manufacturing Engineering
	Engineering Leadership Development	UG	Undergraduate Minor	
	Engineering Entrepreneurship	UG	Undergraduate Minor	
	Humanitarian Engineering and Social Entrepreneurship	UG	General Engineering (GENG) Courses	
	Engineering Design (with Tracks in Systems Design, and in Product Design)	PG	M.Sc. and M.Eng. (more professionally oriented) in Engineering Design	

(continued)

Table 5.1 (continued)

Entity and housing college	Name	Scope	Brief description	Examples of departments in the same institution
Division of Engineering Science Faculty of Applied Science and Engineering —University of Toronto	The Foundations Years Y1 and Y2	Freshman and Sophomore	Two Years Foundational Engineering Programme	Department of Civil Engineering Department of Mechanical and Industrial Engineering Department of Electrical and Computer Engineering
	Praxis II Showcase	UG	Multidisciplinary engineering design for resolving a local challenge in Toronto	
	Engineering Science Majors (Biomedical Systems Engineering, Infrastructure Engineering, Robotic Engineering)	UG	Multidisciplinary engineering majors	
	The Professional Experience Year Programme	UG	One-year Co-Opt Internship Programme with Industry	
	Skoll BASc/MBA Programme	UG/PG	Combined Multidisciplinary Engineering Science B.Sc. and MBA programme in collaboration with School of Management	

School of Engineering Education at Purdue University, which was established in 2004. Other examples are the Department of Engineering Education (since 2004) at Virginia Tech, Centre for Engineering Innovation at Texas A&M (since 2012), and the Department of Engineering and Technology Education at Utah State University (since 2005). The most noticeable emergence outside the USA can be observed in South Korea, where the number of engineering innovation and education focused entities rose from zero, 7 years ago, to 65, currently.

5.3.2 *Brief Description TIEE Structure*

The TIEE aims in particular to bridge the existing gaps in engineering education system of focus on design, innovation and technology startups, pedagogies, engineering practice and higher linkage with industry, as well as well-rounded competencies development. These gaps exist to some extend in Qatar and quite often exist elsewhere in the world, even in the developed countries.

5.3.2.1 TIEE Objectives

The main objectives of TIEE are as follows:

1. **Objective 1**—Provision-competent multidisciplinary/general engineering education programmes and courses.
2. **Objective 2**—Elevate multidisciplinary technology innovation, design, entrepreneurship, and transfer.
3. **Objective 3**—Emphasize students' professional engineering experience, social responsibility, and twenty-first-century engineering competencies development.
4. **Objective 4**—Establish higher order linkages with industry.
5. **Objective 5**—Advance engineering education research, scholarship, practice, and K-12 outreach.

5.3.2.2 Multidisciplinary Structure and Faculty

The TIEE is a multidisciplinary entity or department as opposed to classical disciplinary engineering departments such as electrical or mechanical engineering. Hence, academic programmes/courses overseen or owned by the TIEE are multidisciplinary/general engineering courses/programmes as opposed to disciplinary courses/programmes offered in classical disciplinary departments. Furthermore, the faculty normally hosted/recruited in a TIEE similar entity are multidisciplinary engineering faculty as opposed to disciplinary faculty in the disciplinary departments. TIEE concept faculty would in particular have background in areas such as or related to:

- Engineering design or systems engineering;
- Technology innovation and entrepreneurship;
- Engineering education scholarship/research;
- Technology enhanced learning/information systems.

Curricular and extra-curricular programmes/courses/activities of a TIEE entity would be characterized mainly by:

- Design, innovation, entrepreneurial, and leadership oriented;
- Hands-on, experiential and active learning oriented, with design and projects at the core;
- Significant proportion of industry-linked activities/courses/projects;
- Utilizes state-of-the-art ICT for learning and teaching;
- Practical skills and soft-skills development oriented;

5.3.2.3 Research and Development Characteristics

The R&D activities of a TIEE entity would be mainly characterized with the following:

- Major focus on areas of innovation and engineering education;
- Applied and contextually highly oriented; it is more of a problem-solving and development-oriented than conventional R&D in science-oriented engineering departments/research centres;
- Contextual problems oriented—focus on industry-linked research for solving problems faced in the industrial sector;
- Entrepreneurial oriented, with tech. transfer, start-ups, and commercialization as a main priority.

5.3.2.4 Innovation and Tech. Start-Ups Facilities

Such an entity will manage also state-of-the-art innovation laboratories, studios, and spaces, including (but not limited to) rapid prototyping studios, additive manufacturing laboratories, product design/production laboratories, emerging and disruptive technologies laboratories, learning innovation spaces, tech. start-ups incubator/accelerator, industry-collaboration spaces, and STEM innovation toolkits.

5.3.2.5 Conceptual Models of a TIEE Structure

Figure 5.6 shows a model of academic orientation and tracks of main activities of a TIEE entity positioning in an integrative with depth of other disciplinary

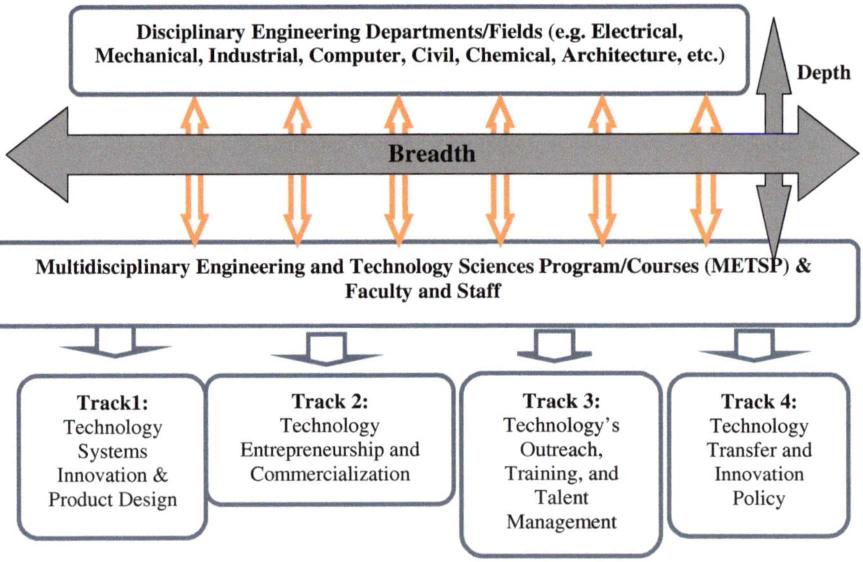

Fig. 5.6 Conceptual model of centre structure and the mutual relationship with other departments in CENG

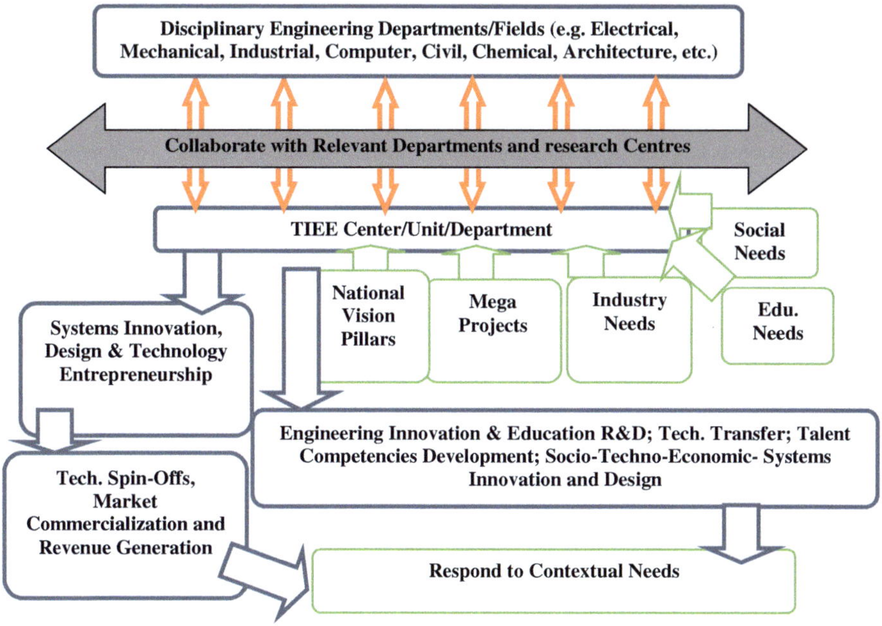

Fig. 5.7 TIEE conceptual model

departments. Figure 5.7 shows a model of a TIEE entity fit within the innovation and iKBE context of a country.

5.4 Conclusions

This chapter describes innovation, design, and technology entrepreneurship in engineering for innovation and knowledge-based economies, "iKBEs," from both global and local perspectives. The technology, innovation, and entrepreneurship triangle as a core of entrepreneurial ecosystem in a iKBE-seeking country was discussed detailing its main three pillars. Further in-depth description of enabling institutional entity in engineering colleges (a so-called Technology Innovation and Engineering Education—TIEE Centre/Department/Unit) is described.

References

Abdulwahed, M., & Hasna, M. O. (2012). Advancing the National Capacity in Engineering/STEM —Concept Framework for Establishing a Center for Engineering Innovation, Education and Entrepreneurship in Support of QNV2030. 40 pp., *QU Proposal*.

Abdulwahed, M., Benammar, M., Hasna, M., & Ghani, S. (2014). Utilizing international design competitions for enhancing 21st century engineering skills. *IEEE Frontiers in Education, Spain, October* (Accepted).

Abdulwahed, M., Gastli, A., Hamouda, A., Ghani, S., & Al-Ammari, R. (2015). Proposal for restructure of the college requirements unit (CRU) into the "Department of Technology Innovation and Engineering Education". 30 pp., Dean's Office, College of Engineering.

Abdulwahed, M., Hasna, M. O., Ghani, S., & Benammar, M. (2016). *Assessment of shell eco-marathon competition impact: Europe, Middle East, and Asia—A Leadership Development Platform*. 121 pp., Industry Research Project Report, Qatar University.

Luo, J., Olechowski, A. L., & Magee, C. L. (2011). *Technologically-based design as a strategy for sustainable economic growth.*

Chapter 6
Methodology, Tools, and Systemic Framework for iKBE Competencies Investigation

6.1 Overview of Methodologies, Target Stakeholders, and Phases of the Study

Various studies have utilized different approaches, tools, and methodologies in their investigations of competencies. Some studies utilized quantitative methodologies only, while other studies utilize qualitative methods only; however, a significant number of reviewed studies utilized a mixed-methods approach whereby both quantitative and qualitative methods are followed. The tools utilized vary from literature views, to surveys, workshops, focus groups, interviews, road maps, etc. Table 6.1 shows an example of studies and their utilized methods and tools; a more extensive list can be found in Appendix C.

The methodology of the assessment included a mixed-methods approach whereby both quantitative and qualitative methods were utilized. The target stakeholders were from both industry and academia. The study was conducted in two main phases: (1)—exploratory and (2)—larger scale.

The following personnel were targeted from industry: (1)—business managers, (2)—training department personnel, (3)—human resources personnel, and (4)—alumni/practicing engineers. In academia, both (1)—senior students and (2)—academics were targeted. Almost, all major engineering disciplines were included in the investigation such as architecture, chemical engineering, civil engineering, computer engineering and science, electrical engineering, industrial engineering, and mechanical engineering. All groups' responses were analysed both qualitatively and quantitatively.

The project is divided into two phases:

- Phase I: Exploratory investigation
- Phase II: Larger-scale investigation

M. Abdulwahed and M.O. Hasna, *Engineering and Technology Talent for Innovation and Knowledge-Based Economies*,
DOI 10.1007/978-3-319-46439-8_6

Table 6.1 Summary of research methodologies

References	Methodology components	Data analysis method (qualitative/quantitative)
USA	Literature review, surveys	Mixed
USA	Workshops, discussions, surveys, focus groups	Mixed
AUS	Survey	Quantitative
Canada	Focus groups	Qualitative
US	Survey, interview, literature review	Mixed

Phase I: This initial phase has been critical in gathering information that later on shaped instruments and planning for the next larger phase of the project. In-depth interviews were conducted with representative personnel of the majority of targeted stakeholders.

Phase II: Instrument surveys were further refined and developed based on exploratory findings in Phase I. Surveys were then distributed for all targeted stakeholders for quantitative data collection.

In the next sections, further details on the utilized qualitative and quantitative methodological approaches are provided.

6.2 Methodologies: Qualitative Approach

For the qualitative part, in-depth semi-structured interviews were held to determine all targeted stakeholders' perceptions. In-depth interviews encourage detailed responses to open-ended questions due to the verbal nature and the ability to have a private or semi-private interaction. Furthermore, the direct interaction permits elaboration on the questions and clarification of any issues. The aim of interviews is to gather qualitative data without concerns on its generalizability; thus, small number of interviews is normally conducted.

6.2.1 Aims of Interviews

The interviews were planned to be the first step in Phase I of the project. Findings were used to fine-tune survey instruments of Phase II as necessary. Interviews are the main step in achieving most of the objectives of Phase I.

6.2.2 *Invitations and Appointments*

Members from industry have been contacted to schedule appointments for the interviews. Selected students and faculty members have been approached within QU campus for the interview.

6.2.3 *Interview Procedures*

Interviews are planned to be semi-structural, face-to-face interactions in the working location of the interviewee, possibly through the phone if the first option was not feasible. Trained interviewers/researchers conducted interviews following an interview protocol. Probing questions were asked whenever necessary to collect more detailed information and to clarify any issues. Respondents' permission was obtained for audio recording of the interviews, transcribing, and thematic analysis.

6.2.4 *Demographics of Qualitative Data*

Qualitative data analysis was carried out according interviews. The total number of interviews is 36 interviews, 7 faculty members, 11 students, and 18 industry representatives (some of the interviews were group interviews).

6.2.4.1 Faculty

For the faculty members' interviews, one from each of the following disciplines was interviewed: chemical engineering, mechanical and industrial engineering, and the architectural and urban planning programme. Also, two faculty members from each of the following were interviewed: computer science and computer engineering, and civil engineering. The majority of faculty members have 18+ years of experience in teaching engineering students and 2+ years in teaching engineering students in Qatar University (QU). All faculty members interviewed have teaching experience outside Qatar/QU, and the majority have working experience in industry.

6.2.4.2 Students

For the students' interviews, one student was interviewed from the chemical engineering programme and two students were interviewed from each of the following programmes: industrial and systems engineering, architectural and urban planning programme, computer engineering, computer science, and electrical engineering. All of the students interviewed are senior students (90+ credit hours finished).

6.2.4.3 Industry

The industry interviews included 3 organizations from the following sectors: construction, research and development, and services. The interviewees were chosen from different departments in the organization which includes managers (senior R&D expert, business development manager, and heads of various departments), HR (HR managers, manager of finance and administration, head of planning and recruitment, and Qatarization planner), training department (senior learning and development advisor for training, head of Qatarization: training and development), and professional engineers or alumni (research and development engineer and project engineer). Most of the interviews done with industry representatives were group interviews.

6.2.5 Interview Questions/Protocols

The interview protocols were developed based on the objective of the study and published interview protocols of other studies in the literature. The introduction of each protocol is intended for the interviewer to introduce him/herself and define the research objectives for the respondent. This was followed by a set of open-ended questions. Each interview protocol had a set of questions classified under different domains as summarized in Table 6.2.

6.2.6 Methodology of Analysis

First, the interviews done were transcribed. Interview transcripts were paraphrased and organized into key points, and the frequency of occurrence of each point was noted. After that, each key point was coded using themes/codes that represent it and key points were rearranged according to theme/code. Then, each key point was checked that it was under the theme/code that best represented it, and finally, key points under each theme were further categorized into subcategories.

Table 6.2 Interview question domains

	Faculty	Senior Students	Managers	Training	Alumni/Practicing	Human resources
Introduction	■	■	■	■	■	■
Background information	■	■	■	■	■	■
Engineering in working place					■	
Graduate competencies	■	■		■	■	
Engineering graduates			■			
Engineering education	■	■	■	■	■	
Engineering students	■					
Quality assurance and linkage to country needs	■					
Future of engineering needs in Qatar			■		■	
Recruitment and employment requirements						■
Job applications						■
Engineering graduates preparedness						■
Links with universities/ industries	■		■	■		■
Internships			■	■		
Concluding remarks	■	■	■	■	■	■

6.3 Methodologies: Quantitative Approach

For the quantitative part, comprehensive surveys (included 20+ questions such as demographics, open-ended, and Likert scale up type questions) have been developed and delivered to the targeted stakeholder groups. Surveys are one of the widely used methodologies in conducting research due to their practicality. Questionnaires are the easiest and the cheapest way to collect data (mainly quantitative) from a large number of respondents, which is the case in this project. Surveys are robust tools of data collection at a scale if they were very well developed and appropriately discriminated and analysed.

6.3.1 Piloting of Surveys

Initial versions of the surveys were piloted on a limited sample number. Feedback was collected from the pilot together with findings from the qualitative interviews and was used to develop the final versions of surveys that were distributed on a larger scale.

6.3.2 Invitations and Surveys for Data Collection Procedure

First, stakeholders were contacted through official email invitation with attached surveys, and meetings were scheduled with researchers to collect data in a semi-structured interview format whereby a stakeholder was asked each question of the instrument and the researchers filled in appropriately. This approach aimed to collect high-quality data and accurate responses from the key stakeholders: industry personnel and academics. For student stakeholders, surveys were distributed at the start of a lecture of a selected course to fill in after short introduction of the researchers for instruments and their purpose. This method is proven to increase the response rate and save time since the responses will be collected with a couple of hours over few days.

6.3.3 Instrument Description

Different questionnaires have been developed to better capture the perspectives of each of the following group of respondents:

- Academia: engineering academics and students.
- Industry: managers, training department, human resources, and alumni/practicing engineers.

Surveys included a number of sections related to the objectives of the investigation in this project. Table 6.3 shows details on different sections for each of the investigated groups; further details can be found in Appendix D.

6.3.4 Reliability and Validity of the Survey

Reliability was demonstrated using **Cronbach's alpha** in order to determine the internal consistencies of the used satisfaction scales; Cronbach's alpha values above 0.9 indicate excellent reliability, and in this study, it was found to be *0.965* for

Table 6.3 Survey question domains

	Faculty	Students	Industry (Managers/Training/HR)	Alumni/ Practicing engineers
Background information				
skills of engineering graduates in Qatar				
Industry-Academia Collaboration				
Undergraduate Engineering Education				
Training, Practice, and employment				
Comments and further suggestions				

students, *0.982* for faculty members, *0.939* for industry, and *0.968* for alumni, indicating high internal consistency.

Validity was also demonstrated through **conclusion validity** which is described as appropriateness of the conclusions reported based on statistical relationships; thus, within this analysis, conclusion validity was reported through the use of inferential statistics relying on statistical significance of the results at the 5 % threshold.

6.3.5 Methodology of Analysis

Descriptive and inferential statistical methods were utilized to analyse the data and report on the findings. Quantitative data analysis was conducted using SPSS (Software Package for Social Sciences). This study makes use of a variety of statistical tools in order to reach its conclusion; about 95 % confidence intervals were produced using both Wilcoxon signed-rank test and Mann–Whitney U test for nonparametric data to determine skills satisfaction and importance gaps between current perception and 2030 one and also between male and female students.

6.3.6 Demographics of Quantitative Data

Surveys were administered to faculty members from Qatar university $N = 41$; also, students were from Qatar university $N = 301$ and junior engineers $N = 28$, and the industry sample number was $N = 35$.

6.3.6.1 Faculty Members

The sample number of faculty members is $N = 41$ in which 95.1 % ($N = 39$) of them were males out of which 5.1 % Qataris and 94.9 % non-Qataris, while females are only 4.9 % of the group analysed. Faculty members who were surveyed belonged to the majority of engineering disciplines including mechanical engineering, industrial and systems engineering, chemical engineering, civil engineering, computer engineering, computer science, electrical engineering, and architecture. The majority of the sample who undertook the survey were experts with average teaching experience of 16.6 years, in which the majority teach senior students.

6.3.6.2 Students

The sample number of students is $N = 301$. The female sample number is $N = 191$ (63.5 %) out of which 35.4 % Qataris and 64.6 % non-Qataris, while males are $N = 110$ (36.5 %) in which 21.1 % Qataris and 78.9 % non-Qataris; total sample number of Qatari students are $N = 90$ (30.2 %) and total sample number of non-Qatari students are $N = 208$ (69.8 %) (with three students who have not replied to this question).

Students who were surveyed belonged to the majority of engineering disciplines including mechanical engineering, industrial and systems engineering, chemical engineering, civil engineering, computer engineering, computer science, electrical engineering, and architecture. The average cumulative GPA of surveyed students is 2.91 out of 4.0, and the average total number of credit hours completed is 116.23.

6.3.6.3 Alumni/Practicing Junior Engineers

The practicing junior engineer sample number is $N = 28$; the majority are males $N = 22$ (78.6 %) out of which 5.6 % Qataris and 94.4 % non-Qataris, while the female sample number is $N = 6$ (21.4 %) with 33.3 % Qataris and 66.7 % non-Qataris. The majority of the sample group have bachelor degrees $N = 26$, and just a few have masters or diploma or other degree.

Most of the surveyed group are currently employed, and it is not their first job since graduation. Most of them (75 %) declared that they got their first employment in 1–6 months after graduation.

The average number of years they have as their job experience is 11.9 years. The employment role for most of them includes project management $N = 14$ and design $N = 8$.

6.3.6.4 Industry Management

Analysis shows that the industrial sector sample number is $N = 35$; the majority of whom are males $N = 28$ (80 %), out of which 7.7 % are Qataris and 92.3 % are non-Qataris. The female sample number is $N = 7$ (20 %) with 28.6 % Qataris and 71.4 % non-Qataris. Industry surveys were collected from three main departments $N = 13$ h, $N = 14$ managers, and $N = 8$ training personnel.

6.4 Systemic Framework for Engineering Competencies Investigation and Implementation

Various visualization models could be considered in order to envisage the tentative picture of the future economy of any country. We propose details of a systems methodological approach that is composed of the following main components or subsystems: (1)—*temporal gap identification system*: current–future; (2)—*situational gap identification system*: academia—engineering graduate—industry; (3)—*temporal control system*: current—tactical—strategic; and (4)—*closed-loop processing system*: input—process—output. The qualitative and quantitative approaches illustrated earlier have been an integral part of the developed systemic approach. The proposed framework can be utilized in consequent studies building upon findings of this investigation in various directions and aspects.

6.4.1 Temporal Gap Identification System

The questions like the industrial area that will evolve, readiness and willingness of the graduates to pick up the skills, and mobility of the engineers to contribute both locally and internationally can be explored for the future. The methodology can start from picking up a nation's path that has moved away from industrial economy to knowledge economy and the talents that they have pursued over a long period of time. This will provide the planners to envisage the traits of engineers in the future (let us say in 2030) and to compare that with the current graduates. For example, if we assume that a certain country wants to achieve the direction that a country X has taken over the years to reach to the current knowledge state, then the country can also study the traits of those leaders who have been able to continuously push the country X to a higher level of knowledge state. Such type of gap analysis (shown in Fig. 6.1) helps to understand the difference between the current state of knowledge and desired state of knowledge.

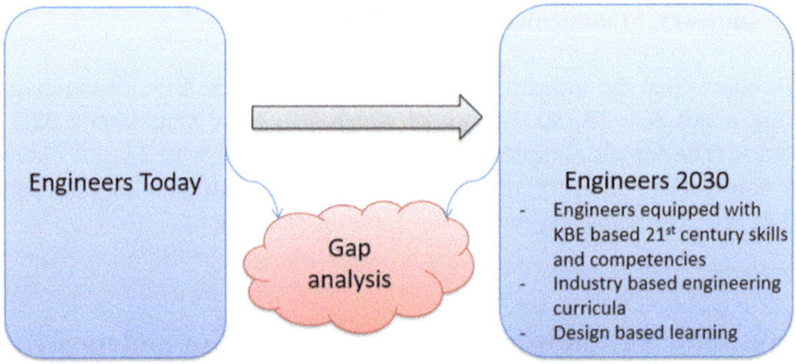

Fig. 6.1 Temporal gap identification system of needed competencies

6.4.2 *Situational Gap Identification System*

A gap analysis can be focused towards employability and competence of graduates, expectations of the industry, society, and businesses, and finally the preparedness of institutions to fill the gaps; *we call this a situational gap analysis*. Therefore, the onus of the framework lies in the changes in the institutions in order to fill the gap over a period of time. Closing a gap is not easy. In Fig. 6.2, five main gap areas are identified based on the current study of engineering practice and the envisaged need for engineers in 2030.

Gap 1—Industry—Graduates' Situational Gap focuses on the industrial needs in terms of competencies in engineers that may relate to the exposure, professional ethics and conduct, understanding of standards and engineers with immersed knowledge on the employer's processes, and strong motivation to contribute and excel in the job.

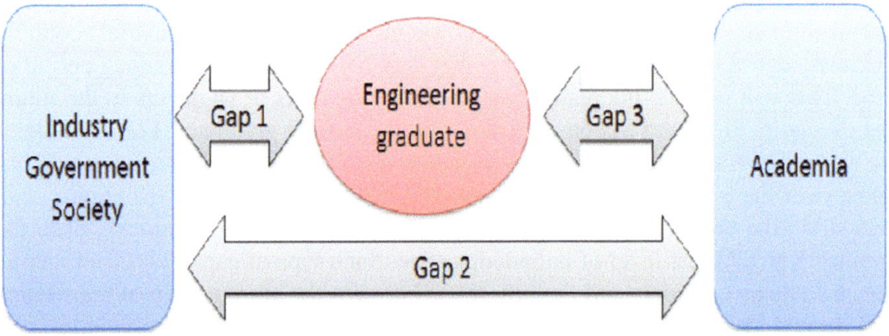

Fig. 6.2 Situational gap identification system

Gap 2—Industry—Academia Situational Gap would highlight the curriculum development, evaluations, and pedagogies in order to meet contextual needs of industry, businesses, and society.

Gap 3—Graduates—Education System Situational Gap would highlight the delivery of curriculum, pedagogies, and evaluation system that develop certain desired competencies in graduate engineers; in other words, it outlines the efficiency and outcome- oriented educational system.

6.4.3 Temporal Control System

The current forecasting method, basically the projection method, may not be sufficient to lead to a desired state of engineers as projection is usually based on the current capabilities and current state of thinking. Instead of forward projection, a backward projection from future state (let us say in 2030) to various years backwards to current level, say 2016 state, can help planners to chart out plans that can be implemented in three levels:

Current (or so-called operational) plan (short-term): that can provide the results within two to three years, through improvements in some part of the curriculum.

Tactical plan (mid-term): that can be implemented within 5 years, for example promoting curriculum changes and mode of delivery such as design-based lecture delivery, and

Strategic plan (long term): for example, introducing industry-based courses at the higher levels, promoting interdisciplinary curriculum and research, introducing sandwich programmes by collaborating with industry, introducing entrepreneurship across the university education, and promoting start-ups and business development for self-employability.

A temporal control system composed of operational–tactical–strategic objectives is shown in Fig. 6.3.

Fig. 6.3 Temporal control system; the system has multi-objectives: operation, tactical, and strategic

6.4.4 Closed-Loop Processing System

The closed-loop system identifies the procedure by which upcoming investigations and implantations emerging from this study will take place. The procedure is composed of three main components, such as shown in Fig. 6.4, (1)—inputs, (2)—process, and (3)—output.

Inputs: define the entrants needed for the procedure. Inputs may be, but not limited to, the following elements: the current level of skills attributes, current frameworks and methodologies for enhancing skills, opportunities for grand challenges, contextual differences in terms of implication of engineering in practice, and finally the potential drivers that lead competence development.

- *Frameworks and Methodologies*: these are the utilized methodologies for investigation in other studies, e.g. mixed-methods approach and scenario-based approach;
- *Grand Challenges*: these are the grand challenges of the country by which engineering and technology can play a major role to meet these challenges;
- *Local and Global Context*: these inputs take into consideration global and local contextual needs and differences; and
- *Change Drivers*: these are the drivers for change in graduate engineers' attributes, e.g. governmental plans and policies, regional and global competition, transformative industry, and development plans.

Process: the information obtained has to be processed in a structured manner, first by scrutinizing data and streamlining them and then by classifying them into various forms and schema as necessary. The data have to be then analysed to see

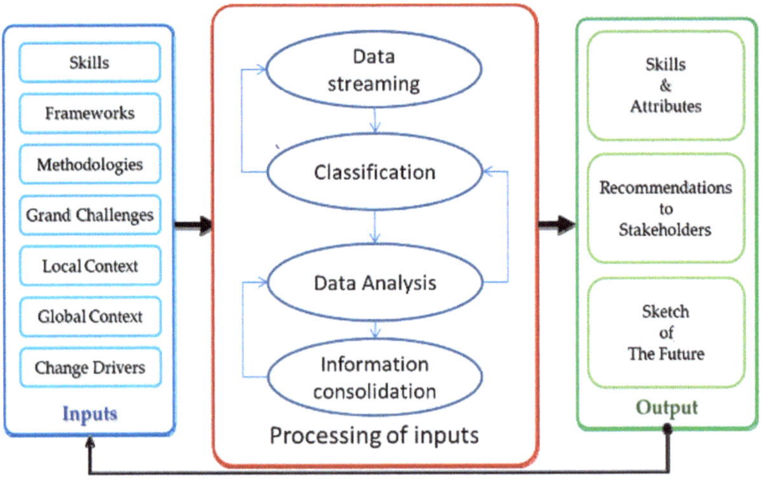

Fig. 6.4 Closed-loop processing system

whether they fit the current and future observations and requirements where views from literature can be of great help. The final step in the processing is the data consolidation in order to develop meaningful resolutions in terms of skills, policies, road maps, and the future paradigm (or sketch of the future). The future paradigm can be developed through the trending method, based on the current processes, or through the back-casting method based on the vision of the future at a particular period.

Outputs define the outcomes extranets of the investigation. The outputs of the processes, as mentioned in the figure, would be a reasonable list of skills, policies for various stakeholders, and the anticipated vision.

- *Skills and Attributes:* focus on competencies needed for engineers to meet the current and future evolving contextual needs;
- *Recommendations:* these are a set of recommendations for adaptation by main stakeholders, e.g. curricular change and governmental policies;
- *Sketch of the Future:* this could be a detailed comprehensive scenario of engineering state and needed industrial transformation for iKBE in a future state (let us say 2030).

The framework is also dynamic in terms of the whole content, in the sense that inputs and processes could be changed with the passage of time. Figure 6.5 shows

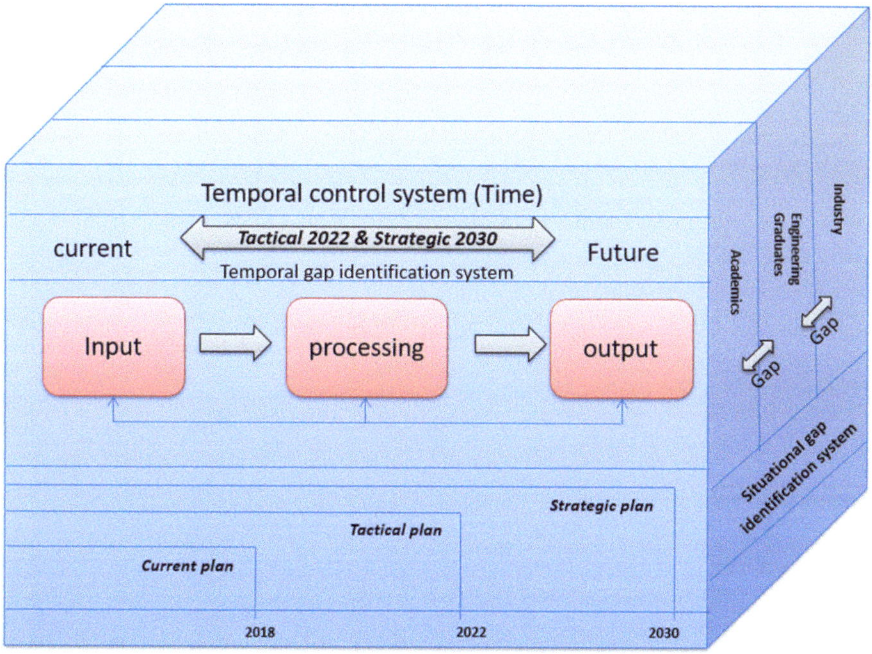

Fig. 6.5 The whole systemic framework with its four subsystems

the whole systemic framework with its four subsystems combined. In the next section, the main findings of the investigation are reported, discussed, and then mapped to the developed systemic framework.

6.5 Conclusions

The methodological approach of constructing the qualitative interview protocols and the quantitative surveys has been detailed. The validity and reliability of the surveys were high as demonstrated by Cronbach's alpha and conclusion validity approaches. Targeted groups of the investigation included stakeholders from industry (managers, trainers, HR, and practicing engineers) and stakeholders from academia (faculty members and senior level students). Analysis of the collected data is provided in the next chapter. The chapter also provided a systemic framework for investigation of needed talent in a national context. The systemic framework is applicable in temporal, situational, and procedural context. The framework will be utilized in the next chapter to synthesize the major findings of the project and develop an implementation road map.

Chapter 7
Empirical National Case Study: Current and Future Engineering Skills Needs, and Engineering Education Systems Necessities in Qatar

7.1 Introduction

The review in Chap. 2 has revealed several national investigations of engineering skills conducted in countries across the world, such as USA, Australia, UK, and Malaysia. A synthesis of skills and competencies was performed and led to the identification of 200+ items. Additionally, content analysis was used to categorize the founded competencies, leading to a set of 22 main items ranked in terms of frequency of repetition in the literature as follows: (1) Communication Competency (CC); (2) Business and Management Competency (PMC); (3) Teamwork Competency (TwC); (4) Problem-solving Competency (PSC); (5) Information and Communication Technology Competency (ICTC); (6) Lifelong Learning Competency (LLLC); (7) Ethics and Responsibility Competency (ERC); (8) Leadership Competency (LsC); (9) Practical Experience Competency (PEC); (10) Critical Thinking Competency (ClTC); (11) Foreign language (FLC); (12) Innovation Competency (InC); (13) Design Competency (SDS); (14) Professionalism Competency (PsC); (15) Decision-making Competency (DMC); (16/17) Inter/Multidisciplinary Competency (I/M-DC); (18) Systems Thinking Competency (STAC); (19) Disciplinary Competency (DC); (20) Entrepreneurship Competency (EpC); and (21) Adabtability Competency (AC); and (22) Math, Physics, and Science Fundamentals Competency (MPSFC). The list shows that there may be a need to reorient the education system, especially when a country aims to achieve innovation-led or iKBE-led economy in the future. This chapter provides empirical analysis on engineering talent and competency needs in Qatar.

M. Abdulwahed and M.O. Hasna, *Engineering and Technology Talent for Innovation and Knowledge-Based Economies*,
DOI 10.1007/978-3-319-46439-8_7

7.2 Analysis of Engineering Skills and Competencies

Stakeholders were asked to evaluate a set of 22 engineering skills on three main levels:

- Skills importance currently.
- Satisfaction level of engineering talent passion of these skills, as an outcome of existing undergraduate engineering education system.
- Skills importance in 2030.

The set of 22 skills are categorized in four dimensions:

- **Dimension I—Core Knowledge and Practice**: disciplinary engineering fundamentals, interdisciplinary engineering knowledge, practical experience, ICT experience, multidisciplinary knowledge, and Science knowledge (math, physics and science fundamentals);
- **Dimension II—Cognition**, **Thinking**, **and Mental**: lifelong learning, problem-solving, decision-making, systems thinking, critical thinking, innovation, and design;
- **Dimension III—Professional** and **Interpersonal**: professionalism, ethics, adaptability, communication, teamwork, and foreign language(s); and
- **Dimension VI—Business and Management**: management, leadership, and entrepreneurship.

Stakeholders were asked to provide their answers on a Likert scale from 1 to 5 where 1 = "strongly disagree", 2 = "disagree", 3 = "neutral", 4 = "agree", and 5 = "strongly agree".

7.3 Stakeholders Qualitative Feedback: Current Skills and Competencies Needs

Stakeholders (industry, faculty, and students) interviewed were asked about competencies and skills engineering graduates required in Qatar. The majority of competencies in the four-dimensional model described earlier in Chap. 2 have been reported by the various stakeholders.

However, frequencies of reports for the following competencies were observed higher than others: communications, teamwork, practical experience, management, lifelong learning, ICT, and disciplinary engineering knowledge. Communications, teamwork, management, and practical experience were indeed on the very top competencies industry required in engineering graduates.

Among the most common weakness/deficiencies in competencies engineering graduates have, the following have been reported: communications, management, practical skills, and foreign language.

7.4 Skills Importance Currently

7.4.1 Industry Versus Faculty

A descriptive analysis showed that both faculty members and industrial sector members means are around 4 "agree," which showed positive perception towards these skills importance currently with relatively higher means in favour of faculty members for **Dimension I**, except for practical experience which industry members showed higher means.

But for **Dimensions II**, **III**, and **IV**, industrial members showed higher means, which was an expected result due to the nature of the faculty members who focus more in the core engineering knowledge listed in Dimension I, while those of the industrial sector focus more on the soft skills and management skills listed in Dimensions II, III, and IV.

Hypothesis testing using **Mann–Whitney U test** showed ***a statistically significant*** difference in ***practical experience*** in favour of industrial sector emphasizing the importance of such skill to the industry as well as ***adaptability***, ***teamwork***, ***foreign language***, and ***management*** as detailed in Tables 7.1, 7.2, 7.3 and 7.4.

Table 7.1 Statistics of 22 generic engineering skills importance currently from S. Students, faculty members, and industrial sector members' perspectives "Dimension I"

Dim I: Core knowledge and practice					
Variable	Group under study	Sample number (*N*)/mean	Mann–Whitney (industry vs. faculty)/mean gap	Mann–Whitney (industry vs. students)/mean gap	Mann–Whitney (faculty vs. students)/mean gap
Disciplinarily engineering fundamentals (depth)	Faculty	41/4.44	0.982/0	***0.033/0.35***	0.080/0.35
	Industry	***61/4.44***			
	S. Students	289/4.09			
Interdisciplinary engineering knowledge (breadth)	Faculty	41/4.22	0.300/−0.17	0.707/0.003	0.374/0.2
	Industry	61/4.05			
	S. Students	283/4.02			
Math, physics, and science fundamentals	Faculty	41/4.22	0.875/−0.07	0.758/0.12	0.659/0.19
	Industry	59/4.15			
	S. Students	281/4.03			
Practical experience	Faculty	41/4.22	***0.033***/0.30	0.241/0.23	0.131/−0.07
	Industry	***60/4.52***			
	S. Students	276/4.29			
ICT experience	Faculty	41/4.41	0.756/−0.05	0.068/0.29	0.151/0.34
	Industry	59/4.36			
	S. Students	286/4.07			
Multidisciplinary knowledge	Faculty	40/3.78	0.827/−0.04	***0.036***/−0.24	0.095/−0.2
	Industry	***61/3.74***			
	S. Students	281/3.98			

Table 7.2 Statistics of 22 generic engineering skills importance currently from S. Students, faculty members, and industrial sector members perspectives "Dimension II"

Dim II: Cognition and thinking					
Variable	Group under study	Sample number (*N*)/mean	Mann–Whitney (industry vs. faculty)	Mann–Whitney (industry vs. students)	Mann–Whitney (faculty vs. students)
Lifelong learning	Faculty	41/4.20	0.053/0.26	0.189/0.23	0.285/−0.03
	Industry	61/4.46			
	S. Students	278/4.23			
Problem-solving	Faculty	41/4.29	0.428/0.14	0.539/0.08	0.156/−0.06
	Industry	61/4.43			
	S. Students	283/4.35			
Decision-making	Faculty	41/3.98	0.060/0.23	0.942/0.02	***0.015***/−0.21
	Industry	62/4.21			
	S. Students	***288/4.19***			
System thinking	Faculty	41/3.95	0.127/0.21	0.714/0	***0.019/−0.21***
	Industry	61/4.16			
	S. Students	***288/4.16***			
Critical thinking	Faculty	41/4.22	0.616/0.06	0.847/0.03	0.389/−0.03
	Industry	61/4.28			
	S. Students	284/4.25			
Innovation	Faculty	41/4.00	0.222/0.26	0.688/0.05	0.078/−0.21
	Industry	62/4.26			
	S. Students	286/4.21			
Design	Faculty	39/4.31	0.923/−0.08	0.364/−0.03	0.262/0.05
	Industry	61/4.23			
	S. Students	284/4.26			

Table 7.3 Statistics of 22 generic engineering skills importance currently from S. Students, faculty members' and industrial sector members perspectives "Dimension III"

Dim III: Professional and interpersonal					
Variable	Group under study	Sample number (*N*)/mean	Mann–Whitney (industry vs. faculty)	Mann–Whitney (industry vs. students)	Mann–Whitney (faculty vs. students)
Professionalism	Faculty	41/4.37	0.399/0.07	0.372/0.18	0.817/0.11
	Industry	62/4.44			
	S. Students	285/4.26			
Ethics	Faculty	41/4.41	0.317/0.11	0.482/0.17	0.579/0.06
	Industry	62/4.52			
	S. Students	284/4.35			
Adaptability	Faculty	41/3.93	***0.011***/0.39	0.349/0.15	***0.028***/−0.24
	Industry	62/4.32			
	S. Students	278/4.17			

(continued)

Table 7.3 (continued)

Dim III: Professional and interpersonal					
Variable	Group under study	Sample number (*N*)/mean	Mann–Whitney (industry vs. faculty)	Mann–Whitney (industry vs. students)	Mann–Whitney (faculty vs. students)
Communication	Faculty	41/4.44	0.164/0.09	0.176/0.22	0.673/0.13
	Industry	62/4.53			
	S. Students	281/4.31			
Teamwork	Faculty	41/4.34	***0.007***/0.27	***0.047***/0.31	0.181/0.04
	Industry	62/4.61			
	S. Students	285/4.30			
Foreign language(s)	Faculty	41/4.29	***0.023***/0.28	0.078/0.31	0.340/0.03
	Industry	60/4.57			
	S. Students	286/4.26			

Table 7.4 Statistics of 22 generic engineering skills importance currently from S. Students, faculty members and industrial sector members perspectives "Dimension IV"

Dim IV: Business and management					
Variable	Group under study	Sample number (*N*)/mean	Mann–Whitney (industry vs. faculty)	Mann–Whitney (industry vs. students)	Mann–Whitney (faculty vs. students)
Management	Faculty	41/3.68	***0.037***/0.37	0.182/−0.10	***0.001***/−0.47
	Industry	62/4.05			
	S. Students	***284/4.15***			
Leadership	Faculty	41/4.00	0.117/0.16	0.617/−0.03	0.016/−0.19
	Industry	62/4.16			
	S. Students	283/4.19			
Entrepreneurship	Faculty	41/3.85	0.844/0	***0.010***/−0.31	0.010/−0.31
	Industry	61/3.85			
	S. Students	279/4.16			

7.4.2 *Industry Versus Students*

Descriptive analysis showed that both senior students and industrial sector members means are around 4 "agree," showing positive perception towards these skills importance currently with relatively higher means in favour of industrial sector members for **Dimension I**, **II**, and **III** except for multidisciplinary knowledge in which senior students showed higher means.

But for **Dimension IV**, skills industry showed higher means which was an expected result due to the nature of the industrial work, which focuses more on the management skills listed in Dimension IV.

Hypothesis testing by application of **Mann–Whitney U test** showed ***a statistically significant*** difference between both groups in ***multidisciplinary knowledge*** and ***entrepreneurship*** **in favour of senior students** and ***disciplinary engineering fundamentals*** and ***foreign language*** **in favour of industry** as detailed in Tables 7.1, 7.2, 7.3 and 7.4.

7.4.3 Faculty Versus Students

Descriptive analysis showed that both faculty members' and senior students' means are around 4 "agree," showing positive attitudes towards these skills importance currently with relatively higher means in favour of faculty members for **Dimensions I and III** skills ***except for adaptability*** which senior students showed higher means. But for **Dimensions II** and **IV**, skills senior students showed higher means than that of the faculty members.

Hypothesis testing using **Mann–Whitney U test** showed ***a statistically significant*** difference between ***adaptability***, ***decision-making***, ***system thinking***, ***leadership***, ***management***, and ***entrepreneurship*** in favour of senior students as detailed in Tables 7.1, 7.2, 7.3 and 7.4.

7.5 Skills Satisfaction

7.5.1 Industry Versus Faculty

Descriptive analysis showed that all means are around 3 "neutral" for both groups showing relatively lower satisfaction degree with these skills although of higher means of faculty members than that of the industrial sector members for both dimensions I and III and higher means for industrial sector members for dimensions II and IV showing higher satisfaction Tables 7.5, 7.6, 7.7 and 7.8.

Also, hypothesis testing using **Mann–Whitney U test** showed ***no statistically significant*** difference between both groups.

7.5.2 Industry Versus Students

Descriptive analysis showed that all means are around 3 "neutral" for both groups showing relatively lower satisfaction degree with such skills. Students have had slightly higher means of satisfaction than Industry for all skills except ICT and disciplinary engineering fundamentals, such as detailed in Tables 7.5, 7.6, 7.7 and 7.8.

Table 7.5 Statistics of 22 generic engineering skills satisfaction from S. Students, faculty members, and industrial sector members' perspectives "Dimension I"

Dim I: Core knowledge and practice					
Variable	Group under study	Sample number (*N*)/mean	Mann–Whitney (industry vs. faculty)	Mann–Whitney (industry vs. students)	Mann–Whitney (faculty vs. students)
Disciplinarily engineering fundamentals (depth)	Faculty	40/3.68	0.599/−0.03	0.323/0.12	0.571/0.15
	Industry	55/3.65			
	S. Students	287/3.53			
Interdisciplinary engineering knowledge (breadth)	Faculty	40/3.45	0.199/−0.23	0.118/−0.24	0.956/−0.01
	Industry	54/3.22			
	S. Students	283/3.46			
Math, physics, and science fundamentals	Faculty	40/3.28	0.072/0.32	0.395/−0.06	***0.010/***0.38
	Industry	53/3.60			
	S. Students	288/3.66			
Practical experience	Faculty	40/3.30	0.407/−0.17	0.169/−0.23	0.578/−0.06
	Industry	53/3.13			
	S. Students	284/3.36			
ICT experience	Faculty	40/3.68	0.533/0.08	0.078/0.29	0.351/0.21
	Industry	54/3.76			
	S. Students	287/3.47			
Multidisciplinary knowledge	Faculty	40/3.33	0.525/−0.2	0.129/−0.27	0.497/−0.07
	Industry	43/3.13			
	S. Students	286/3.40			

Table 7.6 Statistics of 22 generic engineering skills satisfaction from S. Students, faculty members and industrial sector members perspectives "Dimension II"

Dim II: Cognition and thinking					
Variable	Group under study	Sample number (*N*)/mean	Mann–Whitney (industry vs. faculty)	Mann–Whitney (industry vs. students)	Mann–Whitney (faculty vs. students)
Life long learning	Faculty	41/3.44	0.897/0	0.365/0.17	0.260/0.17
	Industry	54/3.44			
	S. Students	287/3.61			
Problem-solving	Faculty	41/3.22	0.744/0.08	***0.005/−0.41***	***0.003/***−0.49
	Industry	53/3.30			
	S. Students	287/3.71			
Decision-making	Faculty	41/3.10	0.442/0.12	***0.034/***−0.31	***0.007/***−0.43
	Industry	55/3.22			
	S. Students	288/3.53			

(continued)

Table 7.6 (continued)

Dim II: Cognition and thinking					
Variable	Group under study	Sample number (*N*)/mean	Mann–Whitney (industry vs. faculty)	Mann–Whitney (industry vs. students)	Mann–Whitney (faculty vs. students)
System thinking	Faculty	40/3.33	0.602/0.11	***0.035***/−0.26	***0.010***/−0.35
	Industry	55/3.42			
	S. Students	286/3.68			
Critical thinking	Faculty	41/3.20	1.000/−0.02	***0.010***/−0.4	***0.015***/−0.38
	Industry	55/3.18			
	S. Students	287/3.58			
Innovation	Faculty	40/3.08	0.781/0.03	0.069/0.32	***0.026***/−0.35
	Industry	55/3.11			
	S. Students	285/3.43			
Design	Faculty	38/3.74	0.051/−0.39	***0.005***/−0.37	0.716/0.02
	Industry	55/3.35			
	S. Students	284/3.72			

Table 7.7 Statistics of 22 generic engineering skills satisfaction from S. Students, faculty members and industrial sector members perspectives "Dimension III"

Dim III: Professional and interpersonal					
Variable	Group under study	Sample number (*N*)/mean	Mann–Whitney (industry vs. faculty)	Mann–Whitney (industry vs. students)	Mann–Whitney (faculty vs. students)
Professionalism	Faculty	41/3.33	0.424/−0.09	***0.007***/−0.12	0.088/−0.03
	Industry	55/3.24			
	S. Students	286/3.36			
Ethics	Faculty	41/3.56	0.481/0.06	0.452/−0.10	0.147/−0.16
	Industry	55/3.62			
	S. Students	284/3.72			
Adaptability	Faculty	41/3.24	0.837/0.01	***0.004***/*−0.41*	***0.004***/*−0.42*
	Industry	55/3.25			
	S. Students	277/3.66			
Communication	Faculty	41/3.63	0.258/−0.29	***0.001***/−0.53	0.071/−0.24
	Industry	56/3.34			
	S. Students	286/3.87			
Teamwork	Faculty	41/3.76	0.734/−0.12	0.207/−0.17	0.367/−0.05
	Industry	56/3.64			
	S. Students	289/3.81			
Foreign language(s)	Faculty	41/3.63	0.691/0.04	0.586/−0.03	0.365/−0.07
	Industry	55/3.67			
	S. Students	289/3.70			

Table 7.8 Statistics of 22 generic engineering skills Satisfaction from S. Students, faculty members and industrial sector members perspectives "Dimension IV"

Dim IV: Business and management					
Variable	Group under study	Sample number (*N*)/mean	Mann–Whitney (industry vs. faculty)	Mann–Whitney (industry vs. students)	Mann–Whitney (faculty vs. students)
Management	Faculty	40/2.98	0.903/0.02	***0.004***/−0.46	***0.004***/−0.48
	Industry	55/3.00			
	S. Students	280/3.46			
Leadership	Faculty	40/3.18	0.981/0	***0.009***/−0.42	***0.006***/−0.42
	Industry	55/3.18			
	S. Students	281/3.60			
Entrepreneurship	Faculty	40/3.85	0.510/−0.85	***0.016***/−0.40	***0.002***/0.45
	Industry	55/3.00			
	S. Students	277/3.40			

Hypothesis testing using **Mann–Whitney U test** showed ***a statistically significant*** difference in most skills as **decision-making**, **problem-solving**, **system thinking**, **critical thinking**, **design skills**, **professionalism**, **adaptability**, **communication skills**, **management**, **leadership**, and **entrepreneurship** ***in favour of senior students***.

7.5.3 *Faculty Versus Students*

Both faculty and students have had an average score around 3 "neutral". Hypothesis testing using **Mann–Whitney U test** showed ***a statistically significant*** difference between both groups for most of the skills such as **math**, **physics**, **and science fundamentals**, **decision-making**, **system thinking**, **critical thinking**, **innovation**, **problem-solving**, **adaptability**, **management leadership**, and **entrepreneurship** in favour of senior students as detailed in Tables 7.5, 7.6, 7.7 and 7.8.

7.6 Skills Importance in the Future

7.6.1 *Industry Versus Faculty*

Descriptive analysis showed that both faculty members' and industrial sector members' means are around 4 "agree," showing their high agreement level towards these skills' importance in the future with relatively higher means in favour of faculty members for **Dimension I**, except for practical experience which industry members have higher means pointing out its importance.

But for **Dimensions II, III,** and **IV**, industrial members showed higher means which was an expected result due to the nature of the faculty members who focus more in the core engineering knowledge listed in Dimension I, while those of the industrial sector focus more on the soft skills and management skills listed in Dimensions II, III, and IV. Generally, both groups have positive attitude towards skills importance by 2030 as detailed in Tables 7.9, 7.10, 7.11 and 7.12.

Hypothesis testing using **Mann–Whitney U test** showed ***no statistically significant*** difference between both groups.

7.6.2 Industry Versus Students

Descriptive analysis showed that both senior students' and industrial sector members' means are around 4 "agree," showing high agreement level towards these skills importance in the future with relatively higher means in favour of industry for all 4 dimensions pointing out skills importance by 2030 to achieve QNV 2030, as detailed in Tables 7.9, 7.10, 7.11 and 7.12.

Table 7.9 Statistics of 22 generic engineering skills importance in the future from S. Students, faculty members, and industrial sector members perspectives "Dimension I"

Dim I: Core knowledge and practice					
Variable	Group under study	Sample number (*N*)/mean	Mann–Whitney (industry vs. faculty)	Mann–Whitney (industry vs. students)	Mann–Whitney (faculty vs. students)
Disciplinarily engineering fundamentals (depth)	Faculty	41/4.59	0.759/0.03	***0.042***/0.32	0.163/0.29
	Industry	60/4.62			
	S. Students	288/4.30			
Interdisciplinary engineering knowledge (breadth)	Faculty	41/4.54	0.937/−0.06	0.063/0.32	0.088/0.38
	Industry	60/4.48			
	S. Students	287/4.16			
Math, physics, and science fundamentals	Faculty	41/4.39	0.699/−0.11	0.311/0.25	0.185/0.36
	Industry	58/4.28			
	S. Students	288/4.03			
Practical experience	Faculty	41/4.46	0.057/0.32	***0.044****/0.35*	0.827/0.03
	Industry	59/4.78			
	S. Students	282/4.43			
ICT experience	Faculty	41/4.41	0.756/−0.05	***0.010****/0.12*	0.073/0.17
	Industry	59/4.36			
	S. Students	288/4.24			
Multidisciplinary knowledge	Faculty	41/4.15	0.938/−0.05	0.552/0.01	0.666/0.06
	Industry	60/4.10			
	S. Students	281/4.09			

Table 7.10 Statistics of 22 generic engineering skills importance in the future from S. Students, faculty members, and industrial sector members' perspectives "Dimension II"

Dim II: Cognition and thinking					
Variable	Group under study	Sample number (*N*)/mean	Mann–Whitney (industry vs. faculty)	Mann–Whitney (industry vs. students)	Mann–Whitney (faculty vs. students)
Lifelong learning	Faculty	41/4.63	0.189/0.09	***0.012***/0.38	0.383/0.29
	Industry	60/4.72			
	S. Students	280/4.34			
Problem-solving	Faculty	41/4.78	0.238/−0.16	0.582/0.21	0.088/0.37
	Industry	60/4.62			
	S. Students	287/4.41			
Decision making	Faculty	41/4.63	0.594/−0.08	0.630/0.17	0.316/0.25
	Industry	62/4.55			
	S. Students	289/4.38			
System thinking	Faculty	41/4.44	0.820/−0.10	0.810/0.03	0.906/0.14
	Industry	61/4.33			
	S. Students	289/4.30			
Critical thinking	Faculty	41/4.68	0.525/−0.12	0.249/0.23	0.095/0.35
	Industry	61/4.56			
	S. Students	290/4.33			
Innovation	Faculty	41/4.51	0.616/−0.10	0.572/0.23	*0.950/0.13*
	Industry	62/4.61			
	S. Students	289/4.38			
Design	Faculty	39/4.56	0.887/0.01	0.837/0.20	0.984/0.19
	Industry	60/4.57			
	S. Students	288/4.37			

Table 7.11 Statistics of 22 generic engineering skills importance in the future from S. Students, faculty members, and industrial sector members' perspectives "Dimension III"

Dim III: Professional and interpersonal					
Variable	Group under study	Sample number (*N*)/mean	Mann–Whitney (industry vs. faculty)	Mann–Whitney (industry vs. students)	Mann–Whitney (faculty vs. students)
Professionalism	Faculty	41/4.68	0.769/0.01	0.082/0.29	0.235/0.28
	Industry	61/4.69			
	S. Students	285/4.40			
Ethics	Faculty	41/4.71	0.666/0.05	0.078/0.33	0.303/0.28
	Industry	62/4.76			
	S. Students	288/4.43			
Adaptability	Faculty	41/4.49	0.564/0.04	0.107/0.27	0.460/0.23
	Industry	62/4.53			
	S. Students	280/4.26			

(continued)

Table 7.11 (continued)

Dim III: Professional and interpersonal					
Variable	Group under study	Sample number (*N*)/mean	Mann–Whitney (industry vs. faculty)	Mann–Whitney (industry vs. students)	Mann–Whitney (faculty vs. students)
Communication	Faculty	41/4.76	0.713/−0.02	***0.016***/0.34	0.090/0.36
	Industry	62/4.74			
	S. Students	284/4.40			
Teamwork	Faculty	40/4.68	0.064/0.13	***0.005***/0.42	0.571/0.29
	Industry	62/4.81			
	S. Students	289/4.39			
Foreign language(s)	Faculty	41/4.71	0.618/0.02	***0.006***/0.43	0.057/0.41
	Industry	60/4.73			
	S. Students	289/4.30			

Table 7.12 Statistics of 22 generic engineering skills importance in the future from students, faculty members, and industrial sector members perspectives "Dimension IV"

Dim IV: Business and management					
Variable	Group under study	Sample number (*N*)/mean	Mann–Whitney (industry vs. faculty)	Mann–Whitney (industry vs. students)	Mann–Whitney (faculty vs. students)
Management	Faculty	61/4.24	0.079/0.27	0.918/0.15	***0.040***/−0.12
	Industry	61/4.51			
	S. Students	288/4.36			
Leadership	Faculty	41/4.41	0.095/0.21	0.207/0.30	0.473/0.09
	Industry	61/4.62			
	S. Students	286/4.32			
Entrepreneurship	Faculty	41/4.37	0.778/−0.07	0.3740	0.583/0.07
	Industry	60/4.30			
	S. Students	285/4.30			

Hypothesis testing using **Mann–Whitney U test** showed ***a statistically significant*** difference in favour of industrial sector as **practical experience**, **ICT skills**, **disciplinary engineering fundamentals**, **lifelong learning**, **communication skills**, **teamwork**, and **foreign language**.

7.6.3 Faculty Versus Students

Descriptive analysis showed that both faculty members' and senior students' means are around 4 "agree," showing high agreement level towards these skills' importance in the future with relatively higher means in favour of faculty members for all dimensions except for ***management skills,*** which students showed higher means as detailed in Tables 7.9, 7.10, 7.11 and 7.12.

Hypothesis testing using **Mann–Whitney U test** showed ***no statistically significant*** difference between both groups.

7.7 Stakeholders Perceptual Gaps in Skills

Gap analysis was carried out in order to know whether there are situational and/or temporal gaps in the opinions; **situational gap** represents the gap between 22 global engineering skills importance currently and satisfaction level, while **temporal gap** represents the gap between skills importance currently and in the future.

7.7.1 Situational Gap: Importance Currently and Satisfaction Level

In order to quantify situational gap, means of satisfaction level was subtracted from the means of skills importance currently. The difference will show that there is a gap between what each group thought is important and how much they are satisfied with it.

**It was assumed that as the difference increases, the gap is more serious, so only skills with highest difference will be highlighted.

7.7.1.1 Faculty Members

For faculty members, the highest skills difference was in practical experience and maths and physics fundamentals in Dimension I; problem-solving, decision-making, critical thinking, and innovation in Dimension II; professionalism, ethics, and teamwork in Dimension III, and entrepreneurship in Dimension IV (see Tables 7.13, 7.14, 7.15 and 7.16).

Table 7.13 Statistics of 22 generic engineering skills situational and temporal gaps "Dimension I"

Dim I: Core knowledge and practice						
Variable	Situational gap/Wilcoxon *p* value			Temporal gap/Wilcoxon *p* value		
	Industry	Students	Faculty	Industry	Students	Faculty
Disciplinarily engineering fundamentals (depth)	0.79/0.032	0.56/0.000	0.76/0.058	0.18/0.000	0.21/0.000	0.15/0.000
Interdisciplinary engineering knowledge (breadth)	0.83/0.000	0.56/0.004	0.77/0.002	0.43/0.000	0.14/0.000	0.32/0.000
Math, physics, and science fundamentals	0.55/0.052	0.37/0.752	0.94/0.070	0.13/0.000	0/0.000	0.17/0.000
Practical experience	1.39/0.005	0.93/0.001	0.92/0.002	0.26/0.000	0.14/0.000	0.39/0.000
ICT experience	0.6/0.002	0.6/0.000	0.73/0.002	0.3/0.000	0.17/0.000	0.25/0.000
Multidisciplinary knowledge	0.61/0.000	0.58/0.021	0.45/0.001	0.36/0.002	0.11/0.000	0.37/0.018

Table 7.14 Statistics of 22 generic engineering skills situational and temporal gaps "Dimension II"

Dim II: Cognition and thinking						
Variable	Situational gap/Wilcoxon *p* value			Temporal gap/Wilcoxon *p* value		
	Industry	Students	Faculty	Industry	Students	Faculty
Lifelong learning	1.02/0.007	0.65/0.023	0.76/0.001	0.26/0.000	0.08/0.000	0.43/0.000
Problem-solving	1.13/0.012	0.5/0.039	1.07/0.000	0.19/0.000	0.2/0.000	0.49/0.000
Decision making	0.99/0.000	0.72/0.000	0.88/0.000	0.34/0.000	0.13/0.000	0.65/0.000
System thinking	0.74/0.001	0.48/0.001	0.62/0.000	0.27/0.000	0.14/0.000	0.49/0.001
Critical thinking	1.1/0.001	0.61/0.057	1.02/0.002	0.28/0.000	0.14/0.000	0.46/0.000
Innovation	1.15/0.000	0.92/0.001	0.92/0.001	0.35/0.000	0.03/0.000	0.51/0.000
Design	0.88/0.001	0.51/0.040	0.57/0.028	0.34/0.000	0.14/0.000	0.25/0.002

Table 7.15 Statistics of 22 generic engineering skills situational and temporal gaps "Dimension III"

Dim III: Professional and Interpersonal						
Variable	Situational gap/Wilcoxon *p* value			Temporal gap/Wilcoxon *p* value		
	Industry	Students	Faculty	Industry	Students	Faculty
Professionalism	1.2/0.001	0.6/0.004	0.93/0.005	0.25/0.000	0.14/0.000	0.31/0.000
Ethics	0.9/0.010	0.63/0.015	0.85/0.042	0.24/0.000	0.08/0.000	0.3/0.000
Adaptability	1.07/0.016	0.51/0.057	0.69/0.000	0.21/0.000	0.09/0.000	0.56/0.000
Communication	1.19/0.002	0.44/0.024	0.81/0.011	0.21/0.000	0.09/0.000	0.32/0.000
Teamwork	0.97/0.001	0.49/0.022	0.58/0.006	0.2/0.000	0.09/0.000	0.34/0.000
Foreign language(s)	0.9/0.075	0.56/0.246	0.66/0.000	0.16/0.000	0.04/0.000	0.42/0.000

Table 7.16 Statistics of 22 generic engineering skills situational and temporal gaps "Dimension IV"

Dim IV: Business and Management						
Variable	Situational gap/Wilcoxon p value			Temporal gap/Wilcoxon p value		
	Industry	Students	Faculty	Industry	Students	Faculty
Management	1.05/0.000	0.69/0.000	0.7/0.000	0.46/0.000	0.21/0.000	0.56/0.000
Leadership	0.98/0.000	0.59/0.005	0.82/0.000	0.46/0.000	0.13/0.000	0.41/0.000
Entrepreneurship	0.85/0.000	0.76/0.002	1/0.000	0.45/0.000	0.14/0.000	0.52/0.000

7.7.1.2 Senior Students

For senior students, the highest skills difference was in practical experience in Dimension I; lifelong learning, decision-making, critical thinking and innovation in Dimension II; professionalism, adaptability, and foreign Languages in Dimension III; and entrepreneurship in Dimension IV (see Tables 7.13, 7.14, 7.15 and 7.16).

7.7.1.3 Industry

And finally for the industrial sector members, the highest skills difference was in practical experience in Dimension I; lifelong learning, problem-solving, critical thinking, and innovation in Dimension II; professionalism, adaptability, and communication in Dimension III; and management in Dimension IV as detailed in Tables 7.13, 7.14, 7.15 and 7.16.

7.7.2 *Temporal Gap: Importance Currently and Importance in the Future*

7.7.2.1 Faculty Members

For faculty members, the highest skills difference was in practical experience in Dimension I; problem-solving, decision-making, critical thinking, and design in Dimension II; professionalism, adaptability, teamwork, and communication in Dimension III; and management in Dimension IV as detailed in Tables 7.13, 7.14, 7.15 and 7.16.

7.7.2.2 Senior Students

For senior students, the highest skills difference was in multidisciplinary knowledge in Dimension I; problem-solving, system thinking, and innovation in Dimension II;

foreign languages, adaptability, and teamwork in Dimension III; and management in Dimension IV as detailed in Tables 7.13, 7.14, 7.15 and 7.16.

7.7.2.3 Industry

For industrial sector members, the highest skills difference was in interdisciplinary knowledge in Dimension I; problem-solving, critical thinking, design, and innovation in Dimension II; professionalism, ethics, adaptability, and communication in Dimension III; and management and leadership in Dimension IV as detailed in Tables 7.13, 7.14, 7.15 and 7.16.

7.8 Change in Skills in 2022 and 2030

7.8.1 Stakeholders Quantitative Perspectives

Stakeholders were asked whether they think that competencies and skills demanded by Qatar industry would change from now till 2022 and from 2022 till 2030. Most respondents replied with "yes" for changes on both time points asked. Overall, the percentage of "yes" was higher for all asked groups for the change long-term point (from 2022 till 2030). Details of groups' answers are shown in Table 7.17.

7.8.2 Stakeholders Qualitative Perspectives

Stakeholders were asked to provide their perspectives on emergences in the engineering profession and competencies needed in Qatar in the future, in particular for the 2022 and 2030 horizons. Stakeholders provided multiple insights on what would potentially be emerging in the future. In sketching the future, stakeholders' main points were as follows:

- Qatar's economy is shifting into a knowledge-based one; the engineering profession and needs in Qatar will be also more innovation and knowledge based;
- Qatar will be producing their own products and increasing their own innovations over the next 5–15 years;

Table 7.17 Percentage of skills change by 2022 and 2030

Group under study	2022/Yes	2030/Yes
Students	79.7 % (N = 208)	85.2 % (N = 219)
Faculty	63.4 % (N = 26)	82.5 % (N = 33)
Industry	62.7 % (N = 43)	76.8 % (N = 43)

- Industry expects higher demands of engineers in the future, and average increase in hiring might be around 15 % a year over the next few years;
- Industry also expects more diverse engineering disciplines to be needed as the country emerges into a iKBE, and more non-traditional disciplines would be needed;
- Industry expected higher demand on recruiting national engineers over the next few years;
- Industry expects higher levels of competencies in leadership and management of next-generation graduate national engineers to cope with high pace of development and transformation;
- There will be need for higher levels of competencies for engineers in the future due to the higher level (in both quantity and quality) of upcoming engineering projects and tasks on ground to be performed by next-generation engineers;
- The transfer into iKBE in the light of QNV2030 implies transformations in the structure and landscape of technology and engineering industry in Qatar, which in turn may imply changes in needs of skills set. In particular, the following skills were reported to be more demanded by various stakeholders: ICT skills, management, leadership, entrepreneurship, innovation, problem-solving, research, practical skills, communications, teamwork, and decision-making.

7.9 Comparative Students Perceptions

7.9.1 Males Versus Females

Generally, descriptive statistics showed that the attitude of males and females is positive for the skills. The means of importance of skills currently and 2030 are around 4 ± which is close to "agree", and the means of satisfaction level are between 3 ± which is close to "neutral" (see Appendix H).

Importance currently: Results showed a statistically significant difference between both groups in the skills importance currently in the following: ***system thinking, critical thinking, innovation, design, adaptability, foreign language, leadership, and decision-making*** importance currently in favour of females (see Table 7.18).

Satisfaction: Females showed overall slightly higher means of satisfaction of skills with higher means than males. However, both groups showed means around 3 "neutral," indicating a lower satisfaction level. A statistically significant difference was revealed only in ***design and professionalism*** (see Table 7.18).

Importance in 2030: A statistically significant difference between both groups in the skills importance in 2030 was revealed for the following: ***interdisciplinary engineering fundamentals (breadth), ICT experience, system thinking, innovation, design, teamwork, and decision-making*** **in favour of females as well** (see Table 7.18).

Table 7.18 Comparative students perception towards generic 22 engineering skills "genders comparison"

Skill	Gender	Mean	*p* value	Skill	Gender	Mean	*p* value
Importance currently				Importance 2030			
Decision-making	M	4.08	0.003	Interdisciplinary knowledge	M	3.96	0.016
	F	4.26			F	4.27	
Systems thinking	M	3.99	0.017	ICT experience	M	4.10	0.047
	F	4.26			F	4.33	
Critical thinking	M	4.07	0.009	Decision-making	M	4.24	0.011
	F	4.35			F	4.46	
Innovation	M	4.00	0.007	System thinking	M	4.14	0.001
	F	4.33			F	4.39	
				Innovation	M	4.29	0.039
Design	M	4.00	0.003		F	4.43	
	F	4.40		Design	M	4.23	0.014
Adaptability	M	3.97	0.017		F	4.45	
	F	4.28		Teamwork	M	4.25	0.009
Foreign language(s)	M	4.08	0.004		F	4.48	
	F	4.37		Satisfaction			
Leadership	M	4.00	0.008	Design	M	3.51	0.026
	F	4.31			F	3.84	
Teamwork	M	4.08		Professionalism	M	3.41	0.003
	F	4.43			F	3.80	

7.9.2 Nationals Versus Non-nationals

Descriptive statistics showed that the attitude of both Qataris and non-Qataris is positive for all skills. The means of importance of skills currently and in 2030 are around 4 ± which is close to "agree", and the means of satisfaction level are between 3 + which is close to "neutral", indicating that they have low satisfaction level with being equipped with such skills (see Appendix I).

Importance currently: Results showed no statistically significant difference between both groups except in ***entrepreneurship*** in favour of non-Qataris (see Table 7.19).

Satisfaction: Qataris showed higher level of satisfaction of such skills with higher means than non-Qataris and no statistically significant difference except in ***multidisciplinary knowledge*** in favour of Qataris (see Table 7.19).

Importance by 2030: Also, results showed no statistically significant difference between both groups except in ***management*** *and l**Leadership*** in favour of non-Qataris as well (see Table 7.19).

Table 7.19 Comparative students perception towards generic 22 engineering skills "nationality comparison"

Skill	Mean	*p* value
Importance currently		
Entrepreneurship		
Qatari	3.99	0.031
Non-Qatari	4.24	
Importance 2030		
Management		
Qatari	4.21	0.042
Non-Qatari	4.43	
Leadership		
Qatari	4.09	0.03
Non-Qatari	4.43	
Satisfaction		
Multidisciplinary knowledge		
Qatari	3.62	0.03
Non-Qatari	3.30	

7.9.3 Ranks of Skills

Stakeholders were asked to rank the top 5 skills in order of importance as stated below. Skills ranking was statistically calculated by the "frequency calculation approach", quantifying the most frequently repeated skill by stakeholders and then calculating its percentage from the total sum of the 22 skills frequency. It is worth mentioning that all stakeholders agreed to rank teamwork and problem-solving from the top 5 skills emphasizing their importance to the engineering graduates.

7.9.3.1 Industry

Rank 1: Communication $N = 33$ (10.9 %);
Rank 2: Disciplinary knowledge $N = 31$ (10.3 %);
Rank 3: Teamwork $N = 26$ (8.6 %);
Rank 4: Ethics/practical experience $N = 22$ (7.3 %); and
Rank 5: Problem-solving $N = 19$ (6.3 %).

7.9.3.2 Students

Rank 1: Ethics $N = 104$ (9.1 %);
Rank 2: Teamwork $N = 103$ (9 %);
Rank 3: Leadership $N = 101$ (8.8 %);
Rank 4: Management/problem-solving $N = 94$ (8.2 %); and
Rank 5: Decision-making $N = 84$ (7.3 %).

7.9.3.3 Faculty Members

Rank 1: Critical thinking $N = 20$ (32.8 %);
Rank 2: Teamwork $N = 19$ (13.1 %);
Rank 3: Problem-solving $N = 16$ (11.5 %);
Rank 4: Communication $N = 15$ (7.9 %); and
Rank 5: Innovation $N = 14$ (7.4 %).

7.10 Gap Analysis and Critical Discussion

7.10.1 Current Most Important Skills for the Engineering Workplace in Qatar

Stakeholders' qualitative feedback reported the particular importance of the following skills set currently: communications, teamwork, practical experience, management, lifelong learning, ICT, and disciplinary engineering knowledge. Quantitative reporting of skills importance in the workplace was obtained via two approaches:

1. Rating on a like-hart scale and
2. Asking the stakeholder groups to write down top important skills needed for the workplace.

Both quantitative approaches revealed a close set of competencies in terms of importance for the workplace (see Tables 7.20 and 7.21).

7.10.1.1 Industry

Industry in general highly values the following set of skills (those appeared in both quantitative approaches): *communications, teamwork, practical experience, disciplinary knowledge, ethics, professionalism, and problem-solving.*

Other important competencies for industry (that appeared in either quantitative or qualitative feedback) include the following: **lifelong learning, professionalism, foreign languages, decision-making, critical thinking, and leadership**.

7.10.1.2 Faculty

Faculty in general highly values the following set of skills (those appeared in both quantitative approaches): *problem-solving, communication, teamwork, ethics, lifelong learning, design, and professionalism.*

Other important competencies for faculty (that appeared in either quantitative or qualitative feedback) include the following: **decision-making, management,**

Table 7.20 Top 7 Skills and competencies importance currently statistical means "like-hart scale"

Top 7 Skills and competencies importance Currently						
Ranking	**Industry**	**Mean**	**Students**	**Mean**	**Faculty**	**Mean**
Top 1	Teamwork	4.61	Problem solving/ ethics	4.35	Communication / Disciplinary knowledge	4.44
Top 2	Foreign language	4.57	Communication	4.31	Ethics/ ICT	4.41
Top 3	Communication	4.53	Teamwork	4.30	Professionalism	4.37
Top 4	Ethics/ practical experience	4.52	Practical experience	4.29	Teamwork	4.34
Top 5	Lifelong Learning	4.46	Foreign language/design/ professionalism	4.26	Design	4.31
Top 6	Professionalism / Disciplinary knowledge	4.44	Critical thinking	4.25	Problem solving	4.29
Top 7	Problem solving	4.43	Lifelong Learning	4.23	Lifelong learning	4.29

Highlights in *yellow* indicate repletion of the competency in the three categories, highlights in *red* indicate repetition in 2 categories, and highlights in *green* indicate only one repletion

Table 7.21 Top 7 skills and competencies importance by the workplace currently in Qatar

Top 7 Skills and competencies importance Currently						
Ranking	**Industry (total Freq. 302)**	**Percentage %**	**Students (total Freq. 1147)**	**Mean**	**Faculty (total Freq. 190)**	**Percentage %**
Top 1	Communication	10.9%	Ethics	9.1%	Critical Thinking	10.5%
Top 2	Disciplinary Knowledge	10.3%	Teamwork	9%	Teamwork	10%
Top 3	Team work	8.6%	Leadership	8.8%	Problem solving	8.4%
Top 4	Practical experience/ Ethics	7.3%	Management / problem solving	8.2%	Communication	7.9%
Top 5	Problem solving	6.3%	Decision making	7.3%	Innovation	7.4%
Top 6	Decision making/ critical thinking/ professionalism	6%	Practical experience	7.1%	Design / Ethics	6.8%
Top 7	Leadership	5.6%	Critical thinking	6.4%	Professionalism / LLL	5.8%

Highlights in *yellow* indicate repletion of the competency in the three categories, highlights in *red* indicate repetition in 2 categories, and highlights in *green* indicate only one repletion

leadership, professionalism, ICT, disciplinary knowledge, critical thinking, and innovation.

7.10.1.3 Students

Students in general highly value the following set of skills (those appeared in both quantitative approaches): *teamwork, ethics, problem-solving, and practical experience.*

Other important competencies for students (that appeared in either quantitative or qualitative feedback) include the following: **lifelong learning, decision-making, professionalism, communication, critical thinking, design, foreign language management, and leadership.**

7.10.2 Communalities Among Groups

There have been certain competencies that have been highly stressed by two or three of the stakeholders groups, which are as follows (see also Fig. 7.1):

- Teamwork, ethics, and problem-solving have appeared in all groups in both quantitative approaches;
- Communication, lifelong learning, professionalism, decision-making, leadership, and critical thinking have appeared in all groups in either quantitative approach.

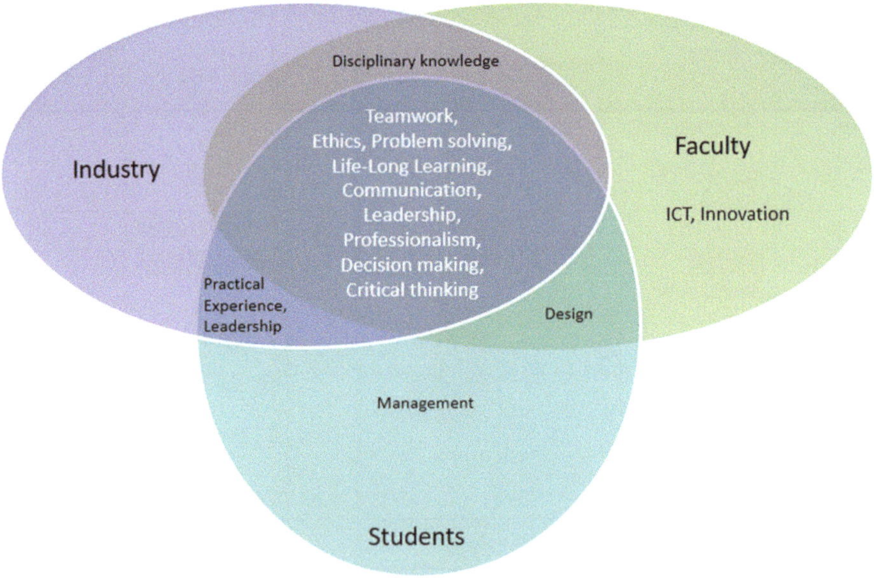

Fig. 7.1 Most important skills for the engineering workplace currently in Qatar, and commonalities and differences among stakeholders groups: (industry, students, and faculty)

7.10.3 Differences Among Groups

A number of differentiating observations have been noticed among different groups, which are as follows (see also Fig. 7.1):

- It is worth mentioning that "practical experience" never appeared among top important competencies that faculty reported in either of both quantitative approaches, while "practical experience" appeared in both industry and students, and in both quantitative approaches. This could be related to the fact that the majority of faculty members are academically and theoretically oriented and have had scarce practice in industry during their career paths.
- Also, industry emphasizes significantly the importance of disciplinary knowledge which appeared only once by either faculty in both quantitative approaches and never appeared by students.
- Both industry and students emphasized the leadership skills' importance being reported once by either of the quantitative approaches, but on the other hand, only faculty members emphasized the importance of ICT and innovation skills which also appeared once by either of the quantitative approaches.
- It is worth mentioning that important competencies for iKBE ready engineers such as innovation, creativity, and design never appeared among top interests currently for the industry sector stakeholders. This could be attributed to the fact that the majority of engineering functions in Qatar at this stage are not much related to iKBE.
- Both faculty and students emphasized the importance of "design", in both quantitative approaches for faculty and once by either of the qualitative approaches by the students; other iKBE competencies such as creativity, innovation (appeared only once by faculty), entrepreneurship, and even management skills (appeared only once by students) have never appeared or barely appeared in the top competencies of industry in both quantitative approaches.

7.10.4 Current Most Gaps in Skills of Fresh Graduates

Stakeholders' qualitative feedback reported the following competencies with significant weaknesses in fresh engineering graduates: communications, management, practical skills, and foreign language.

In quantitative analysis overall, all situational gaps (except two for students) for all stakeholders (industry, faculty, students) were on the positive side, indicating a gap between satisfaction of engineering graduates' competencies and importance of these competencies in the workplace. All situational gaps recorded a statistically significant difference; the gaps were higher for industry and faculty than those of students. In comparing satisfaction level of both academics and industry of engineering talent, it was found that stratification is rather low and there is no

Table 7.22 Highest situational gaps reported by the all stakeholder groups

Highest 7 Skills and competencies Satisfaction			
Ranking	**Industry**	**Students**	**Faculty**
Top 1	Practical experience (1.39)	Practical experience (0.93)	Problem solving (1.07)
Top 2	Professionalism (1.20)	Innovation (0.92)	Critical thinking (1.02)
Top 3	Communication (1.19)	Entrepreneurship (0.76)	Math, Physics and sciences (0.94)
Top 4	Innovation (1.15)	Decision making (0.72)	Professionalism (0.93)
Top 5	Problem solving (1.13)	Management (0.69)	Innovation (0.92)
Top 6	Critical thinking (1.10)	Lifelong Learning (0.65)	practical exp. (0.92)
Top 7	Adaptability (1.07)	Ethics (0.63)	Decision making

Highlights in *yellow* indicate repletion of the competency in the three categories, highlights in *red* indicate repetition in 2 categories, and highlights in *green* indicate only one repletion

statistically significant difference; both industry and academia are less satisfied about competencies than students' own satisfaction with several competencies that had statistically significant lower mean for faculty and industry than that of students.

The highest situational gaps reported by the all stakeholder groups were in the following set of competencies: practical experience, innovation, professionalism, problem-solving, communications, critical thinking, decision-making, math/physics/sciences, lifelong learning, management, and entrepreneurship (see Table 7.22 for more details).

Interestingly, stakeholders of the three groups (industry, faculty, and students) reported the lowest satisfaction levels on competencies highly correlated with a iKBE ready engineer, such as innovation, entrepreneurship, management, leadership (and related attributed such as critical thinking and decision-making), problem-solving, multidisciplinary knowledge, interdisciplinary knowledge, and practical experience (see Table 7.23 for further details).

7.10.5 *Future Trends of Skills Demands in Qatar*

The majority of stakeholders (industry, faculty, and students) expected changes in the demand of competencies set of engineering graduates in Qatar in the future (both for 2022 and for 2030). In particular, qualitative feedback from stakeholders highlighted the following set to be of higher demand: more demanded by various stakeholders—ICT skills, management, leadership, entrepreneurship, innovation,

Table 7.23 Lowest satisfaction levels on competencies and skills

Least 7 Skills and competencies Satisfaction						
Ranking	Industry	Mean	Students	Mean	Faculty	Mean
Least 1	Management	3.00	Practical experience	3.36	Management	2.98
Least 2	Entrepreneurship	3.00	Professionalism	3.36	Entrepreneurship	3.00
Least 3	Innovation	3.11	Entrepreneurship	3.40	Innovation	3.08
Least 4	Practical experience	3.13	Multidisciplinary knowledge	3.40	Decision making	3.10
Least 5	Multidisciplinary knowledge	3.13	Innovation	3.43	Leadership	3.18
Least 6	Leadership	3.18	Management	3.46	Critical thinking	3.20
Least 7	Critical thinking	3.18	Interdisciplinary knowledge	3.46	Problem solving	3.22

Highlights in *yellow* indicate repletion of the competency in the three categories, highlights in *red* indicate repetition in 2 categories, and highlights in *green* indicate only one repletion

problem-solving, research, practical skills, communications, teamwork, and decision-making.

In quantitative analysis, overall, all temporal gaps for all stakeholders (industry, faculty, and students) were on the positive side, indicating a higher importance or demand of engineering graduates competencies in the future (2030) than currently. All temporal gaps recorded a statistically significant difference, and the gaps were relatively higher for the expert group (industry and faculty) than those reported by students. Industry in general scored higher means than faculty (faculty scored higher than students) on the future (2030) importance of practical skills, and on the majority of Dimension 2 (cognition, thinking, and mental), Dimension 3 (professional and interpersonal skills), and Dimension 4 (business and management); however, no statistically significant difference was detected between faculty and industry.

It is worth mentioning that analysis of the top temporal gaps reported by the expert group (industry and faculty) has shown tendency to value the importance of iKBE-oriented competencies higher in the future than other evaluated skills (temporal gaps were top highest for these competencies). Expert group in particular reported highest temporal gaps in highly correlated iKBE competencies such as innovation, entrepreneurship, leadership (and related attributes such as decision-making, adaptability, and system thinking), management, interdisciplinary knowledge, multidisciplinary knowledge, design, and problem-solving (see Table 7.24 for further details).

Table 7.24 Stakeholders temporal gaps

Temporal gap	
Industry	**Faculty**
Management (0.46)	Decision making (0.65)
Leadership (0.46)	management (0.56)
Entrepreneurship (0.45)	Adaptability (0.56)
Interdisciplinary (0.43)	Entrepreneurship (0.52)
Multidisciplinary (0.36)	Innovation (0.51)
Innovation (0.35)	Problem solving (0.49)
Design/ Decision making (0.34)	System thinking (0.49)

Highlights in *yellow* for repletion in two categories

7.11 Qualitative Feedback: Challenges and Barriers

7.11.1 Challenges in Joining the Workplace for Fresh Graduates

Stakeholders (industry, faculty, and students) throughout the interviews have reported multiple challenges that could be faced by fresh graduates, which have been categorized as follows:

- **Lack of experience**: For the vacancies in the market, companies hire engineers from outside because of the experience, and in many cases, they could accept a lower salary. Companies' expectations of fresh graduates are not met since they look for experience when hiring. Preference is given to recruits with 5+ years of experience, and also, there is preference for GCC experience. Skills set shortage with national fresh graduates is normally compensated through structured, comprehensive, and relatively long training programmes (sometimes can last 2 years); these training programmes are not available for non-national fresh graduates.
- **Gender issues**: Some companies (especially small companies) do not prefer hiring female engineers, for matters related with sustainability; for instance, in case of maternity leave, there will be need of a substitute and extra overhead paid. Some females may not prefer working in certain engineering environments (e.g. offshore). Females are usually, based on their own preference, appointed to work in planning and design (office work).

- **Environment**: Diversity of cultures in working environment, ethical matters, conflicts, reality versus theoretical world in college, etc.; long working hours/shifts, offshore work, tough environmental conditions, etc.
- **Skills shortage**: Deficiency in skills (both technical and non-technical), such as communication skills, teamwork skills, andprofessionalism. The knowledge that the students obtained in the universities is not enough, so they need to learn independently and stay updated.
- **Motivation**: Several stakeholders in particular from the industry reported that motivation for working in real engineering tasks is rather low with national fresh graduate (in particular, males), as they tend to aspire for management positions as soon as possible.
- **Nationality**: Non-national fresh graduates may have significant difficulties as compared with nationals, and they have to compete with overseas recruits who have more experience and can be employed for less.
- **Engineers shortage and competition**: Qatar industry has a significant shortage of engineers, in particular skilled ones and national engineers, which has been frequently reported. Competing on national engineering talent is very high among industry employers in Qatar.

7.11.2 Barriers from Preparing Students with the Desired Skills

Industry, faculty, and students reported several barriers from preparing students with desired skills, and the two main emerging themes were as follows:

- **Motivation**: Students normally tend to achieve the minimum requested. The students are normally grade oriented but not learning oriented. Overall speaking, males are less motivated than females and nationals are less motivated than non-nationals.
- **Lack of significant practical experience in engineering curriculum**: This concern has been frequently raised by all stakeholders groups (industry, faculty, and students). Engineering curriculum is more theory oriented, and less focus is given to the professional side of engineering. Internships are normally 8 weeks long, conducted in the summer, and are not enough. Furthermore, the vast majority of faculty members do not have industry experience (very similar to majority of engineering sciences schools worldwide).

Table 7.25 Descriptive statistics of actions that help to prepare students for work

Variable	*N*	Mean	Median	Mode
The university changes the curriculum depending on the needs of the country in the future	41	3.88	4	4
Increase collaboration with industry	*40*	*4.53*	*5*	*5*
Sector-specific internship/work placements as an integral part of the engineering study programme	39	4.08	4	5
Reward faculty members who conduct engineering education research	41	3.76	4	5
Faculty members think that there should be a professional practice course	39	3.74	4	4
Need of providing professional development office	41	3.76	4	4

7.12 Academic Perspectives on Enhancing Graduates' Readiness for Industry Employment

Faculty members in engineering have been asked about their agreement level that certain actions by engineering education institutions will help better prepare graduates for the work place in a scale-up with a scale ranged from 1 to 5 where 1 = strongly disagree, 2 = disagree, 3 = neutral, 4 = agree, and 5 = strongly agree. A number of items were stated as shown in Table 7.25, and these items were deducted based on the previous in-depth interviews with selected faculty members. Faculty responses were positively skewed towards agreement above the "neutral" point for all items, and the highest mean of faculty responses was recorded the item on "*increase collaboration with industry*". In general, the surveyed faculty supports further adaptation of the engineering curriculum for country needs, increasing internships, and rewarding of faculty members who conducts engineering education research.

7.13 Learning and Teaching Styles for Practice Readiness: Students/Junior Engineers Perspectives

Students and alumni were asked about the learning teaching styles/approaches that are useful in preparing/qualifying students for the engineering practice in a Likert scale-up question rated in a scale ranged from 1 to 5 where 1 = "strongly disagree", 2 = "disagree", 3 = "neutral", 4 = "agree", and 5 = "strongly agree". Learning and teaching styles/approaches included 16 different methods and were spread over four dimensions:

- **Classical methods (more passive oriented)**: lectures, tutorial sessions, individual assignment, and written examination;

- **Feedback and support methods**: instructor office hours, instructor feedback on performance, and help and guidance from teaching assistants;
- **Constructivist methods (more active oriented)**: group study, research/design projects, group assignments, oral presentations, class discussions, and online resources; and
- **Practical methods**: internships/practical experience, practical examinations, and laboratories.

7.13.1 Students Perceptions: Males, Females, Nationals, and Non-nationals

Descriptive statistics showed that students agreed that most of the teaching styles mentioned are useful for the engineering practice with means close to 4 "agree" for both males and females. In general, females have had slightly higher means than males for most of the asked styles; however, a statistically significant difference was detected for two styles only: group assignments and instructor office hours. From nationality perspective, there was non-statistically significant difference in styles between nationals and non-nationals.

7.13.2 Senior Students Versus Junior Engineers

Descriptive statistics showed that junior engineers have higher perceptions than senior students for the majority of asked styles. Senior students' perception was lower only for two styles (no statistically significant difference): "instructor office hours" and "tutorial sessions". A statistically significant difference was found in favour of junior engineers for the following styles: group study, internship training experience, oral presentation, practical examinations, and written examinations (see Table 7.26).

7.13.3 Internship

Students and junior engineers were asked whether they were involved in internship or training experience during their undergraduate education, and if yes, they were asked whether this experience was beneficial or not.

About 60 % of both the students (N = 159) and junior engineers (N = 17) pointed that they have internships before. Most of the students 86 % (N = 127) and junior engineers 81 % (N = 13) stated that it was beneficial.

Table 7.26 Statistics of the students and junior engineers' evaluation of teaching styles received during their university study

Teaching style	Variable	Mean		(*N*)	*p* value	Sig.
Classical methods	Lectures	Jun. Eng.	3.96	28	0.173	No
		Sen. Stud.	3.61	273		
	Tutorial sessions	Jun. Eng.	3.50	28	0.284	No
		Sen. Stud.	3.65	269		
	Individual assignment	Jun. Eng.	3.79	28	0.207	No
		Sen. Stud.	3.45	271		
	Written examinations	*Jun. Eng.*	*4.04*	28	*0.001*	*Yes*
		Sen. Stud.	3.23	271		
Feedback and support methods	Instructor office hours	Jun. Eng.	3.54	28	0.491	No
		Sen. Stud.	3.60	272		
	Instructor feedback on performance	Jun. Eng.	4.15	27	0.600	No
		Sen. Stud.	3.94	269		
	Help and guidance from teaching assistances	Jun. Eng.	4.00	28	0.456	No
		Sen. Stud.	3.77	270		
Constructivist methods	Group study	*Jun. Eng.*	*4.18*	28	*0.003*	*Yes*
		Sen. Stud.	3.50	272		
	Research/design projects	Jun. Eng.	4.39	28	0.078	No
		Sen. Stud.	3.99	270		
	Group assignment	Jun. Eng.	3.93	28	0.337	No
		Sen. Stud.	3.64	274		
	Oral presentations	*Jun. Eng.*	*4.41*	27	*0.012*	*Yes*
		Sen. Stud.	3.81	273		
	Class discussion	Jun. Eng.	4.48	27	0.012	No
		Sen. Stud.	3.90	270		
	Online learning resources	Jun. Eng.	3.79	28	0.941	No
		Sen. Stud.	3.68	267		
Practical methods	Internshipand training experience	*Jun. Eng.*	*4.64*	28	*0.039*	*Yes*
		Sen. Stud.	4.07	269		
	Practical examinations	*Jun. Eng.*	*4.29*	28	*0.015*	*Yes*
		Sen. Stud.	3.66	273		
	Laboratories	Jun. Eng.	4.07	27	0.525	No
		Sen. Stud.	3.89	271		

7.13.4 Skills and Competencies Gained from Specific Courses and Practical Training: Qualitative Feedback

Students were asked to highlight the most useful experiences they have during their undergraduate studies. The most frequent experiences and teaching learning styles significantly praised by students were as follows: senior/design projects, group projects, laboratories, practical training, presentations, competitions, and volunteering. Students also highlighted the importance and pedagogical and teaching and learning approach of instructors. Practical courses/styles and extracurricular activities were reported as a main platform for developing competencies by students, while just a few mentioned theoretical courses.

7.13.5 Discussion on Curriculum for Better Workplace Readiness

Qualitative feedback from stakeholders reported two major themes of preparing students with desired skills for workplace: **lack of practical aspects/experience in engineering education curriculum** and **low motivation of students**. Faculty highlighted several means in which engineering graduates' readiness for workplace can be enhanced, and in particular, the following three approaches scored the highest means: increased collaboration with industry, increased internships and practical training, and changes in the curriculum to further meet the country's contextual needs. Most of the senior students and junior engineers who had an internship during their studies agreed that it was beneficial means for work readiness preparation. Both the senior students and junior engineers reported highest means on experiential, practical, and active learning and teaching approaches, when they asked on best approaches they experienced to prepare them for work readiness. Internship and practical training came at number 1, and other approaches included research/design projects, laboratories, practical examinations, oral presentations, and classroom discussions (see Table 7.26 for further details).

It is worth mentioning that such approaches (internships, design projects, laboratories, presentations, etc.) contribute significantly for developing competencies such as practical experience, communications, teamwork, problem-solving, professionalism, which in particular has been highly regarded by industry as shown earlier.

7.14 Academia–Industry Collaboration

7.14.1 Importance of Academia–Industry Linkage

Faculty members and industry were asked how much they perceive in collaboration with industry/academia to be beneficial/important for both sides in a scale-up question ranged from 1 to 5 where 5 = very important, 4 = important, 3 = not important, 2 = not important at all, and 1 = N/A.

Table 7.27 Industry–academic collaboration statistical analysis

Variable	Sample no. (*N*)		Means	Mann–Whitney	Statistical significance
Industry–academia collaboration	Industry	32	4.72	0.040	Yes
	Faculty	39	4.36		

Faculty mean response was close to 4 "important" (4.36). Industry mean response was higher mean (4.72) close to "very important", indicating higher awareness of the importance of industry–academia with statistical significance between both groups in favour of industrial sector with p value ≤ 0.005 after using **Mann–Whitney U test**, as shown in Table 7.27.

7.14.2 Means of Collaborations

Faculty members and industry were asked whether their organizations have industry–academia linkages/collaborations. They were also asked to specify means of collaboration and the aspects in which they would like to collaborate in the future. A number of collaboration aspects/means were stated in the survey (total of seven aspects), together with a self-reporting dimension under "Other" items; these are shown in Table 7.28.

Table 7.28 Statistics of industry–academia collaboration currently and in the future

Variable	Industry–academia collaboration currently				Industry–academia collaboration in the future			
	Sample no. (*N*)		Yes	Percentage (%)	Sample no. (*N*)		Yes	Percentage (%)
Engineering and career events	Industry	32	16	50	Industry	25	12	48
	Faculty	37	26	70.3	Faculty	28	20	71.4
Internships to graduates.	Industry	32	22	68.8	Industry	25	16	64
	Faculty	35	26	70.3	Faculty	29	18	62.1
Through "real-life" projects and resources	Industry	31	20	64.5	Industry	24	15	62.5
	Faculty	34	17	50	Faculty	28	17	60.7
University provides off-site workforce training for employee	Industry	32	17	53.1	Industry	25	12	48
	Faculty	36	29	80.6	Faculty	28	18	64.3
Others	Industry	31	9	29	Industry	25	7	28
	Faculty	34	18	52.9	Faculty	28	16	57.1

This was a multiple response question so sum of percentages may exceed 100 %

The majority of faculty members (89.5 %, $N = 34$) have reported collaborations with industry in one or more forms. The means of collaboration currently include the following: university provides off-site workforce training $N = 29$ (80.6 %), internships to graduates and engineering and career events $N = 26$ (70.3 %), and real-life projects $N = 17$ (50 %).

While means of collaboration in the future include offering internships to graduates $N = 16$ (64 %) and real-life projects $N = 15$ (62.55), university provides off-site workforce training for employee and engineering career event $N = 12$ (48 %).

Majority of industrial members have reported collaborations with academia $N = 29$ (87.9 %). The means of collaboration currently include offering internships to graduates $N = 22$ (68.8 %) and providing "real-life" projects and resources $N = 20$ (64.5 %), university provides off-site workforce training $N = 17$ (53.1 %) and engineering and career events $N = 16$ (50 %).

While means of collaboration in the future includes engineering and career events $N = 20$ (71.4 %), university provides off-site workforce training for our employee $N = 18$ (64.3 %)and offers internships to graduates $N = 18$ (62.1 %) and real-life projects $n = 17$ (60.7 %).

7.14.3 Barriers of Industry–Academia Collaboration

Faculty members and industry were asked whether they think that there are barriers that limit interaction with industry/academia, and also if they agree, they were asked to show their agreement on the barriers suggested in a scale-up question ranged from 1 to 5 where 1 = strongly disagree, 2 = disagree, 3 = neutral, 4 = agree, and 5 = strongly agree.

Descriptive statistics showed that the majority of faculty members $N = 27$ (71.1 %) thought that **there are barriers** that limit interaction with industry on the other hand of those of the industrial sector who do not believe so $N = 9$ (28.1 %).

Faculty members ranked the top 3 major barriers for collaboration as follows: (1) lack of organizational/structural office/unit to coordination $N = 18$ (63.4 %), (2) lack of autonomy to work with industry $N = 15$ (53.6 %), and (3) the process is too complicated and has no collaboration from industry side $N = 14$ (50 %), while industry representatives rated the major 2 barriers for collaboration as follows: lack of organizational/structural office/unit to coordination $N = 3$ (44.4 %) and the university structure is not adapted to the needs of industrial collaborations $N = 3$ (33.3 %), as detailed in Table 7.29.

Table 7.29 Statistics of industry–academia collaboration barriers

Variable	Industry–academia collaboration barriers			
	Sample no. (*N*)		"Yes"	Percentage (%)
Inadequate infrastructure	Industry	9	2	22.2
	Faculty	28	12	42.9
Lack of autonomy to work with industry	Industry	9	2	22.2
	Faculty	28	15	53.6
The university structure is not adapted to the needs of industrial collaborations	Industry	9	3	33.3
	Faculty	28	10	35.7
The collaboration is time-consuming	Industry	9	2	22.2
	Faculty	28	10	35.7
Lack of organizational/structural office/unit to coordination	Industry	9	4	44.4
	Faculty	28	18	64.3
There is no value or benefit to our university/industry from such collaboration	Industry	9	1	11.1
	Faculty	28	2	7.1
We had a bad experience with industry/university in the past	Industry	9	0	0
	Faculty	28	0	0
The process is too complicated	Industry	9	2	22.2
	Faculty	28	14	50
No collaboration from industry/university side	Industry	9	2	22.2
	Faculty	28	14	50
Others	Industry	9	2	22.2
	Faculty	28	3	10.7

This was a multiple response question so sum of percentages may exceed 100 %

7.15 Qualitative Perspectives on Industry–Academia Linkage

The majority of interviewed stakeholders (industry and faculty) highlighted how important it is for the educational institutions to have links and collaborations with the industry. Several **reasons for the importance** of **collaboration** with industry which includes have been reported as follows:

- Students will have a better chance of practical training;
- Graduates will have higher chances of getting accepted for work;
- Industries will give a feedback on the quality of its graduates, and the university will be aware of the modifications they need to do on the curriculum to suit the industry needs;
- It enhances the image of the university by sending a message that the university has qualified professors and students;
- It will enable faculty to get more insight on the industry, in particular for those who does not have industry experience;

- It will enable higher collaboration between academia and industry in terms of R&D, and knowledge and technology transfer; and
- Overall, it enhances the engineering education experience and curriculum.

Several faculties reported that they have already some links with industry, in particular through senior design projects or research projects. Faculty members also suggested a number of ways in which linkages with industry can be established or enhanced, such as (1) joint conferences/seminars/lectures; (2) senior projects related to industry; (3) research projects; (4) establishment of a dedicated office; (5) field visits; (6) sponsorship/internships of students; (7) industry chair positions.

It is also worth mentioning that industry has reported that QU could also offer industry consultations, training, certificate, etc. in professional emerging areas in their sectors based on assets and expertise of existing faculty. For instance, some of the proposed areas were in management, energy systems, risks, etc.

7.16 Towards 2030: Systemic Framework Mapping of Project Findings and Proposed National Roadmap

7.16.1 Temporal Gap Identification System: Current–Future

Temporal gap identification system in Fig. 7.2 illustrates the skills required by Qatar state to move away from hydrocarbon-based economy to innovation and knowledge-based economy. It highlights the engineering talents and traits that, when well understood, will bridge the difference between the current state of knowledge and desired state of knowledge. Temporal gaps were previously highlighted in Chap. 5. Emphasized skills and attributes are those needed for current graduates in Qatar to be equipped with.

7.16.2 Situational Gaps Identification System: Academia–Engineering Graduates–Industry

7.16.2.1 Gap# 1 Industry—Engineering Graduates Gap

Both industrial sector members and engineering graduates revealed 4 major gaps: (1) employability, (2) competencies level of engineering graduates, (3) practical experience, and (4) motivation.

Employability Employability :as a gap has two folds: (1) industrial sector complained about the lack of skilled fresh graduate national workforce capable of providing high-quality work and (2) the non-nationals have some problems in

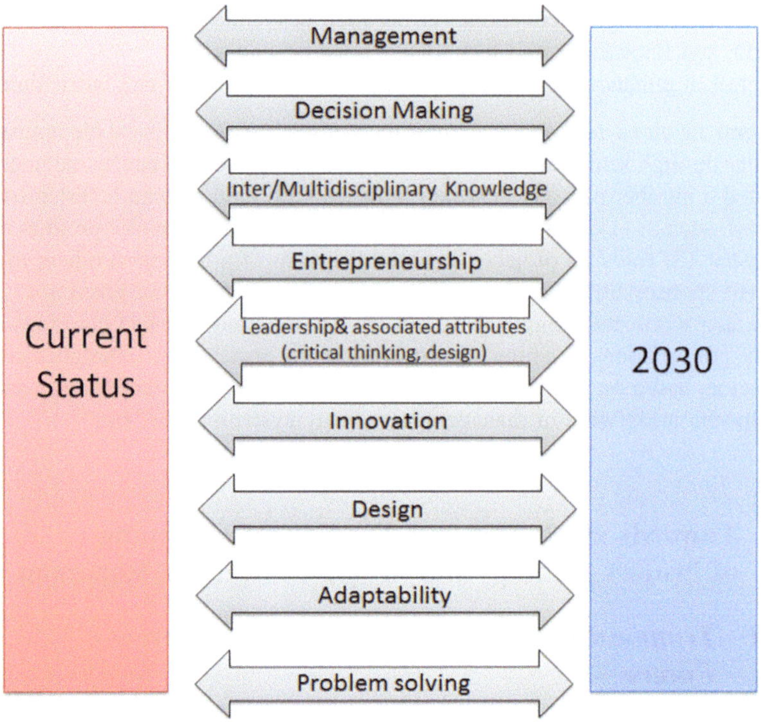

Fig. 7.2 Temporal gap identification system

finding suitable work after graduation due to lack of practical experience and competition with experienced overseas recruits.

Competencies level of engineering graduates: There have been two folds identified in skills gaps. First, industry highlighted significant gaps in top workplace competencies such as practical experience, professionalism, communications, and teamwork; second, satisfaction levels were minimal in regard to competencies related to iKBEs talent attributes such as innovation, entrepreneurship, and leadership (see Fig. 7.3 for an illustration of the two-folded situational gaps in competency level of engineering graduates).

Practical experience: Practical experience was a major issue identified by industry in terms of gap.

Motivation: Motivation was highlighted by industry as one of the main themes that showed a statistical significance between both males and females in favour of females and non-being highly motivated than those of the nationals; also, other universities' graduates have higher motivation but not statistically significant than QU graduates.

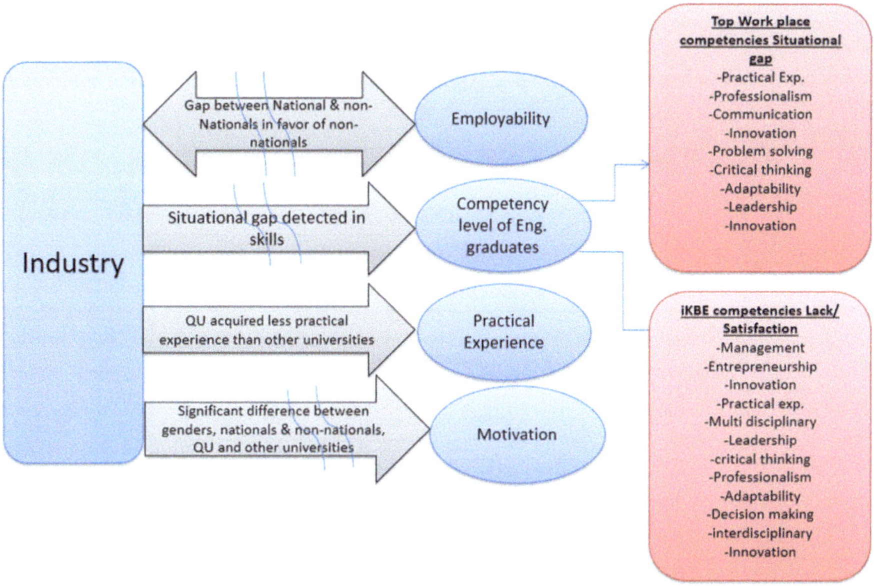

Fig. 7.3 Industry–engineering graduates situational gap

7.16.2.2 Gap # 2 Industry–Academia Gap

The industrial sector and academia showed so many gaps related to 3 main themes: (1) curriculum gaps, (2) perceptual importance gap, and (3) collaboration structure (see Fig. 7.4).

Curriculum gaps: To realize Qatar's vision of providing full potential, highly qualified workforce, students must be able to move easily between education curricula and workplace requirements. No longer can education pathways be considered a linear stand-alone progression without mutual collaboration between industrial sectors and institutions, so industry-integrated curricula was needed. Also, industry pointed out the presence of curriculum gaps mainly in practical experience which they pointed the importance of having practice-integrated model of education to meet the skill requirements of a more diversified economy.

Perceptual importance gap: The interest of both academia and industry in having highly qualified graduates equipped with a set of skills and competencies showed some gaps as per the interest of each group whereby academia's main interest was in the theoretical educational knowledge, while those of industry focus more on the practical hands-on skills. Gaps in practical skills were highlighted significantly by industry, indicating dissatisfaction with current hands-on experience of recent graduates. Also, innovation and design skills importance highlighted by the academia but not emphasized by those of the industry.

Collaboration structures: A statistically significant gap was revealed in the level of importance of having academia–industry collaboration. The industrial

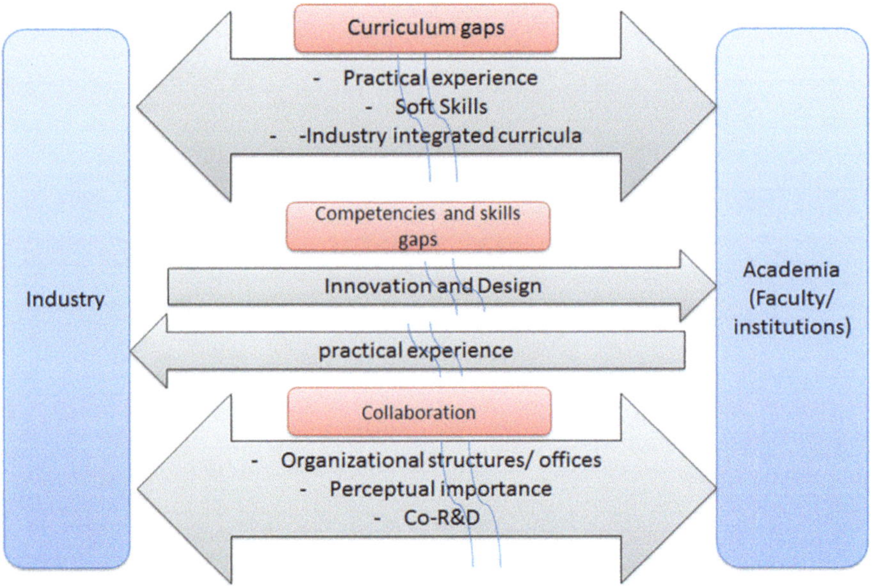

Fig. 7.4 Industry–academia situational gap

sector emphasized the importance of collaboration more than those in academia. Both groups pointed out the lack of organizational/structural office/unit to coordinate joint activities. Industry emphasized more on the university structure that was not adapted to the needs of industrial collaborations and faculty members emphasized more on lack of autonomy to work with industry pointing out the need of Co-R&D.

7.16.2.3 Gap # 3 Engineering Graduates–Academia Gaps

Industry and faculty members' perception towards students showed many gaps that were mainly highlighted when they were asked to compare males versus females, nationals versus non-nationals, and Qatar University versus other universities in Qatar (see Fig. 7.5).

Genders comparison: Industry and/or faculty members stated that females are better than males in terms of motivation, academic performance, and GPA, being more organized with better planning and management skills and so more prepared to field work.

Nationality comparison: Overall, there was a gap in terms of nationality identified, and non-nationals perceived to achieve academically higher. This might be attributed to the fact that overall, they are admitted on more competitive basis and those selected are already having relatively higher competencies levels collectively.

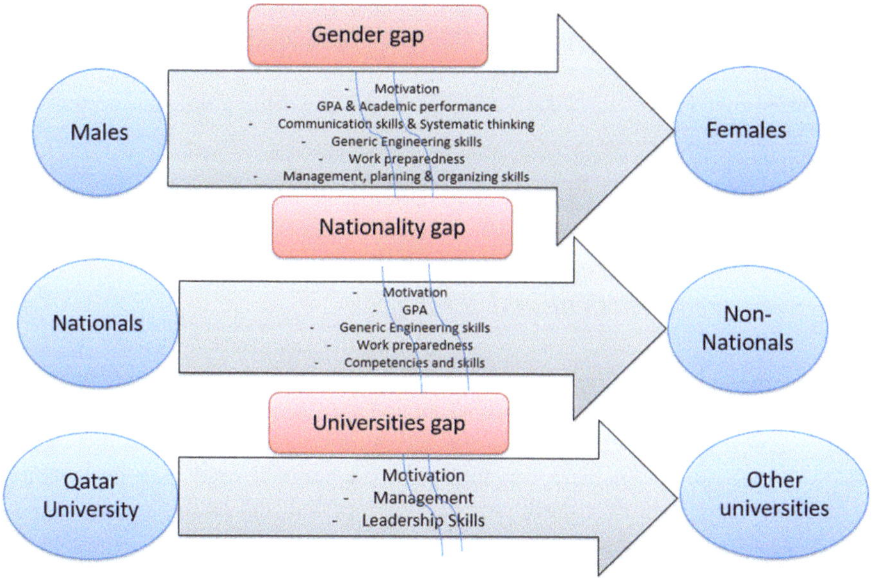

Fig. 7.5 Engineering graduates–academia situational gap

Universities comparison: Faculty members stated that other universities' graduates are better equipped with management and leadership skills in addition of being more motivated. Qatar university graduates were perceived to have higher theoretical and technical knowledge than graduates of other universities in Qatar.

7.16.2.4 Gap Identifications: The Implications

There are a number of implications that can be inducted from the gaps identifications and has been conducted in this study, which are as follows:

1. Engineering education institutions in Qatar should focus more on developing engineering graduates with innovation and knowledge-based competencies in order to meet 2030 emerging needs in the country;
2. Overall, engineering education institutions in Qatar should enhance the competency levels of current graduates to further meet the workplace needs;
3. Clearer employability tracks should be investigated for non-national engineering graduates in Qatar, in close communications and collaboration between the academia and industry sectors;
4. Practical and industry-based experience should be more integrated in engineering and technology curricula of engineering schools in Qatar;

5. Further investigations on motivation aspects of engineering graduates (root analysis and interventions testing) should be conducted in Qatar;
6. The engineering curricula in Qatar should become more adaptive/personalized in order to address the identified gaps in engineering graduates in terms of gender and nationality.

7.16.3 *Temporal Strategic Control System: Current-Tactical-and Strategic*

Based on the project findings, identified gaps, literature reviews, contextual analysis, and various consolidated observations and feedbacks, we propose a temporal strategic control system described in the subsections below, as shown in Fig. 7.6. The proposed framework here can function as a national road map for enabling higher order and level of contributions of engineering and technology in driving innovation and knowledge-based economy in the state of Qatar.

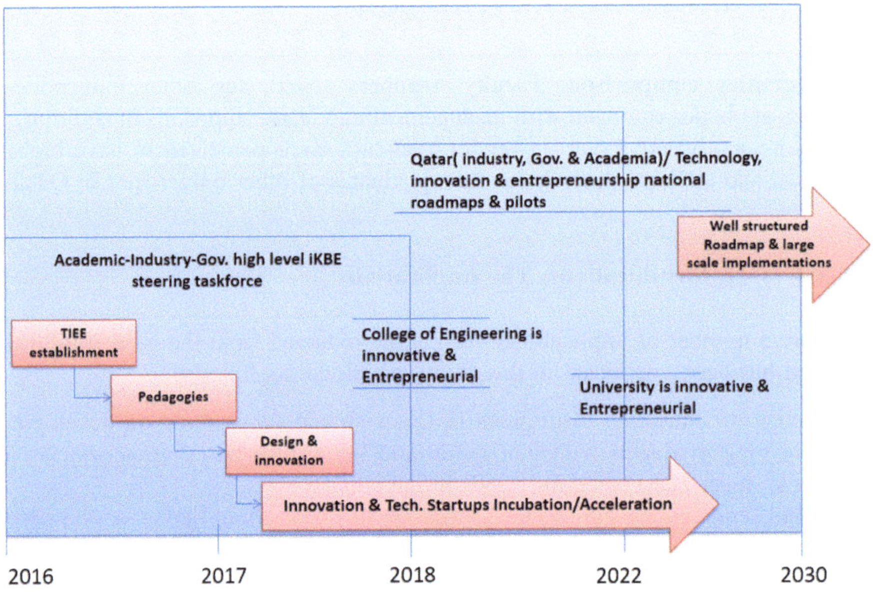

Fig. 7.6 Temporal control system

7.16.3.1 Current Interventions, Outcomes, and Implementations

Establishing Technology Innovation, Entrepreneurship, and Education (TIEE): Establishing TIEE organizational structures in existing engineering education institutions in Qatar is highly recommended. The main duties of these structures are to bridge gaps identified in engineering graduates and work as an effective organizational interface between academia and industry. Such structures will also introduce and scale up innovative engineering pedagogies in engineering curricula in Qatar, and drive technological innovation, engineering design, and technology start-ups from academia, transforming existing engineering colleges into innovation and entrepreneurial entities. Some emerging technology start-ups from academia are expected by the end of this stage.

Establishing a High Level National Academia–Industry–Government Tech. Innovation Steering Taskforce: The proposed taskforce will enable higher level of communications, coordination, development, and strategic implementations of a *National Tech. Innovation and Entrepreneurship Roadmap* across the three major enabling sectors (government, industry, and academia), or what we call the tech. innov. and enter. triangle, building upon existing assets and programmes.

7.16.3.2 Tactical Interventions, Outcomes, and Implementations

In this stage, a consensus over a *National Tech. Innovation and Entrepreneurship Roadmap* has been achieved and several pilot implementations are taking place across all involved key sectors (government, industry, and academia). Significant increase in government-enabling programmes and incentives, financial-enabling ecosystem, industrial transformations into iKBE activities, tech. start-ups, tech. transfer, patenting, industry–academia joint projects, etc. is expected during this stage.

7.16.3.3 Strategic Interventions, Outcomes, and Implementations

In this stage, the *National Tech. Innovation and Entrepreneurship Roadmap* is being implemented on a large scale. The aforementioned expectations in the tactical stage are envisaged to grow in scale, quality, and economic-added value in this stage.

7.16.4 Closed-Loop System: Input–Process–Output

We propose a closed-loop system in which developments, pilots, implementations, roadmaps, recommendations, programmes, etc. emerge from the proposed National

Tech. Innovation Steering Taskforce and the National Tech. Innovation and Entrepreneurship Roadmap to be reported, monitored, evaluated, and revised on periodic basis. From systems and control theory perspectives, the closed-loop control approach will enable higher robustness of outcomes against emerging uncertainties and will keep developments in track with agreed on objectives and visions.

7.17 Conclusions

This chapter provides empirical analysis on engineering talent competency needs in Qatar as a case study. Targeted stakeholders (industry, academia, and students) were asked to evaluate a set of 22 engineering skills on three main levels: importance currently, in the future, and satisfaction. Communications, teamwork, management, and practical experience were indeed on the very top competency industry required in engineering graduates although of being the least satisfied in engineering graduates as reported by both qualitative and quantitative methodological analysis. Skills situational gaps and hypothesis testing proved that students have higher own satisfaction with several competencies although of being unsatisfactory with both of industry and of academia. Skills temporal gaps have shown higher tendency to value the importance of skills and competencies to Qatar`s future. Respondent from all stakeholders expected changes in the demand of competencies set of engineering graduates in Qatar in the future for both 2022 and 2030 horizons. Furthermore, stakeholders' perspectives on industry–academia collaboration have been provided in detail. Finally, findings of the empirical investigation were mapped into the systematic framework developed earlier in the methodology chapter.

Chapter 8
Highlights of Foundational Principles of Engineering & Technology Innovation and Talent Transformation for iKBEs Development

8.1 The Engineer of iKBE: Aspects of Aspirational Engineering Education System

In this section, we provide a set of recommendations for implementations for engineering institutions in an iKBE-seeking country based on the overall findings and critical discussions in this study. These recommendations are generally applicable for MENA institutions and elsewhere where a knowledge-based economy is sought. The proposed set of recommendations spans over four dimensions: 1—context and industry, 2—innovation and venturing, 3—curriculum and structure, and 4—pedagogy and scholarship:

- Context and Industry
  - Contextualize engineering education for the local needs while keeping global flavour,
  - Foster inherent linkage with industry through multiple channels.
- Innovation and Venturing
  - Foster multidisciplinary design and innovation,
  - Focus on technology entrepreneurship in engineering education,
  - Focus on technology leadership in engineering education.
- Curriculum and Structure
  - Deploy an integrated engineering education curriculum with focus on engineering practice, theoretical knowledge, and technical and soft competencies development,
  - Develop and deploy structured pathways for engineering faculty development in engineering innovation and education,

M. Abdulwahed and M.O. Hasna, *Engineering and Technology Talent for Innovation and Knowledge-Based Economies*,
DOI 10.1007/978-3-319-46439-8_8

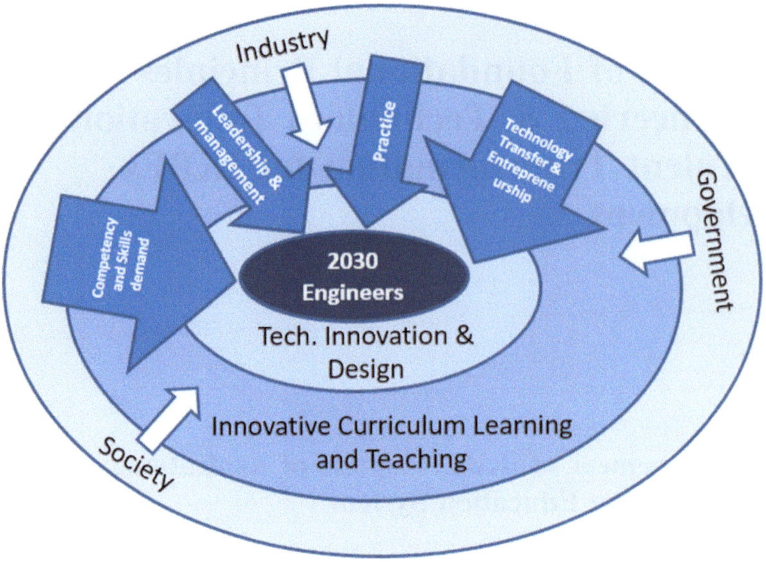

Fig. 8.1 Aspects of aspirational engineering education system in an iKBE-seeking country

- Develop and deploy driving organizational structures in form of a centre or department for engineering innovation and education (e.g. the TIEE concept).

- Pedagogy and Scholarship
  - Foster constructivist pedagogy and experiential approaches in engineering education,
  - Utilize recent innovations in technology enhanced and digital learning in engineering education.

The engineering and technology educational system in an iKBE-seeking country will have the future engineer as a core, with innovation and design curricular layer, innovative learning and teaching layer, and a curriculum that is highly responsive and integrated with the industry, society, and government needs. We recommend four distinguishing further emphasizes dimensions, which are:

1. Leadership and management,
2. Engineering practice,
3. Technology transfer and entrepreneurship,
4. Soft skills competencies development beyond the technical.

Figure 8.1 shows a conceptual model of an aspirational engineering education system in an iKBE-seeking country.

8.2 The TIEE Concept: An Implementation Arm for iKBE Engineering Education Systems

One of the main highlights of the book is the model of an organizational structure to be implemented in engineering education intuitions to better enable the following:

1. Focus on multidisciplinary innovation, design, and entrepreneurship,
2. Soft skills development,
3. Interface with industry and solicitation of contextual needs in order to map into core engineering curriculum,
4. Elevate learning and teaching innovations and scholarship.

The model of a TIEE (Technology Innovation and Engineering Education/ Entrepreneurship) concept has been explained in further depth in this chapter and is visualized again here in Fig. 8.2.

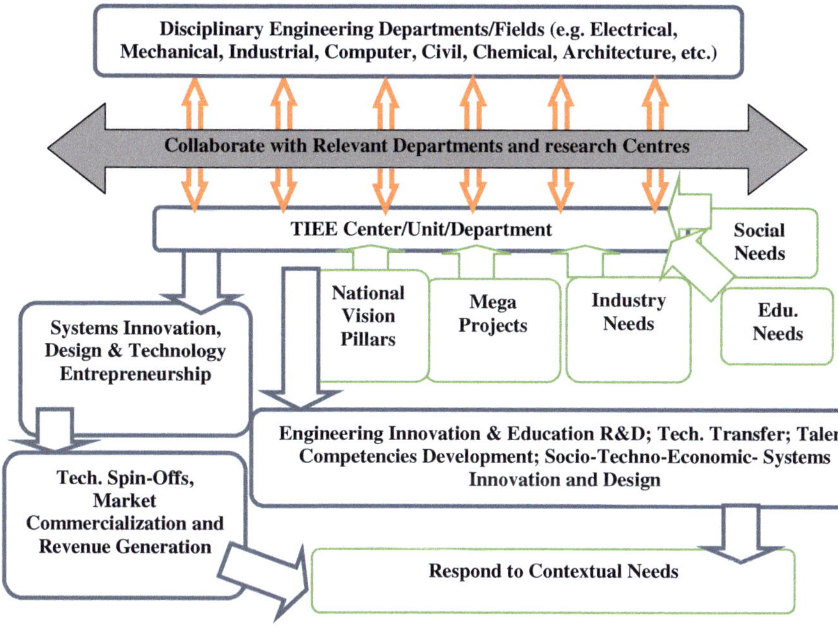

Fig. 8.2 TIEE conceptual model

8.3 Needs of Investigations on the Competencies in Higher Levels of the Pyramid

The project focused mainly on the base level (level 1 and some aspects of level 2) of the competencies pyramid model developed earlier in Chap. 2 and visualized again here in Fig. 8.3. Further in-depth analysis and projects will be needed to investigate competencies demands and set on the subsequent levels of the pyramid: 1—the domain-specific level, 2—the subdomain-specific level, and 3—the functional-specific level.

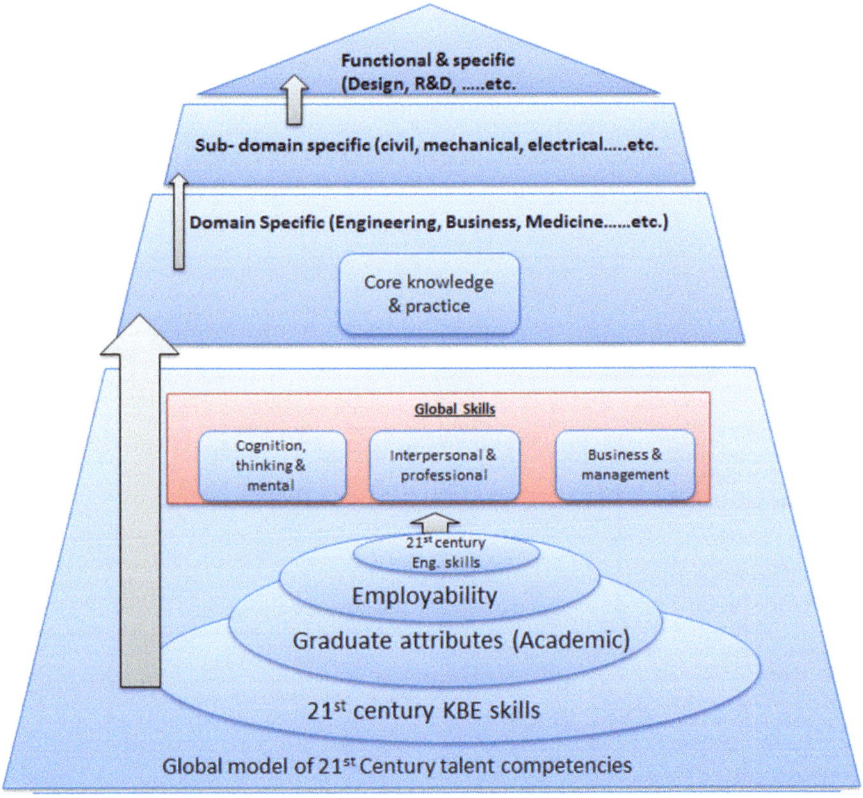

Fig. 8.3 Conceptual framework of the relation among skills (from generic to specific)

8.4 Beyond the Sole of Academia on the Road to 2030: The Technology, Innovation, and Entrepreneurship Triangle

Obviously, engineering academic intuitions should aim to produce the best fit of Engineers 2030 for the country needs. However, in addition to developing competent knowledge-based engineering education systems for 2030 in the academic side, we believe that a innovation and knowledge-based economy (iKBE) can be led and grown rapidly through focus on integrating academia, industry, and government in a systematic implementation road of technology, innovation, and entrepreneurship ecosystem.

Hence, one of the major themes emerging as a recommendation from the project is to form a high-level strategic national innovation steering committee composed of members from the three sectors. Initial conceptualization of road maps to emerge from such committee has been discussed earlier in the previous section and has been visualized in Fig. 8.4.

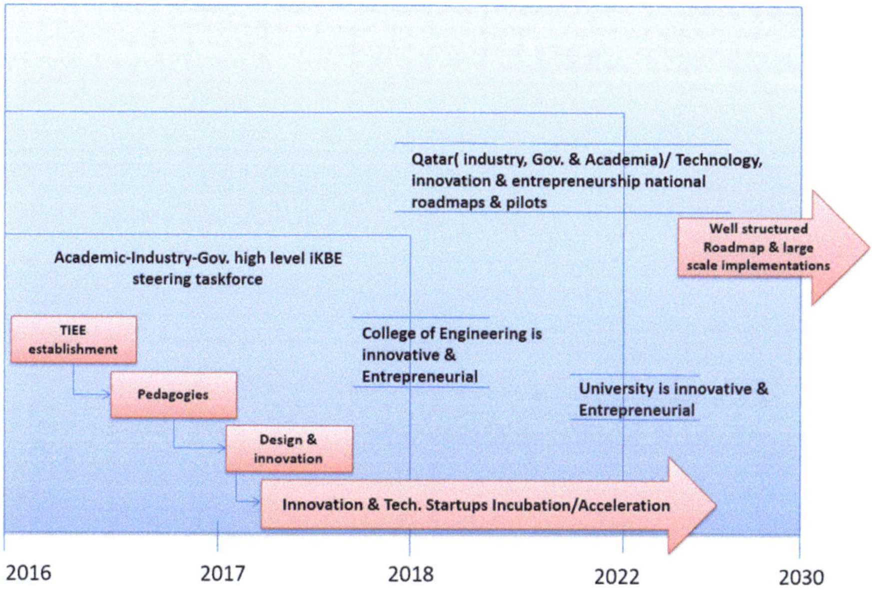

Fig. 8.4 Initial conceptualization of an innovation steering committee road map

8.5 Conclusions

This chapter provided main highlights of the project and the next steps further to build upon its findings. There were four main themes involving conceptual models: 1—aspirational education system for iKBE, 2—implementation arm for engineering education systems, 3—future investigation on the upper levels of the competencies pyramid, and 4—beyond the sole of academia, towards an integration of industry–government–academia in a technology, innovation, and entrepreneurship triangle for meeting iKBE transformation.

Appendix A
Engineering Education Accreditation Criteria

ABET Engineering Criteria 2000 (USA)

The following student outcomes along with any additional outcomes articulated by the programme:

- An ability to apply knowledge of mathematics, science, and engineering.
- An ability to design and conduct experiments, as well as to analyse and interpret data.
- An ability to design a system, component, or process to meet desired needs within realistic constraints such as economic, environmental, social, political, ethical, health and safety, manufacturability, and sustainability.
- An ability to function on multidisciplinary teams.
- An ability to identify, formulate, and solve engineering problems.
- An understanding of professional and ethical responsibility.
- An ability to communicate effectively.
- The broad education necessary to understand the impact of engineering solutions in a global, economic, environmental, and societal context.
- A recognition of the need for, and an ability to engage in lifelong learning.
- A knowledge of contemporary issues.
- An ability to use the techniques, skills, and modern engineering tools necessary for engineering practice.

OSC Engineering Occupational Standards (UK)

- Develop engineering products and processes.
- Produce engineering products and processes.
- Install engineering products and processes.
- Operate engineering products and processes.
- Improve the quality and safety of engineering products and processes.
- Plan, implement, and manage engineering projects.
- Develop own engineering competence.
- Develop own engineering competence (Dodridge 2001).

M. Abdulwahed and M.O. Hasna, *Engineering and Technology Talent for Innovation and Knowledge-Based Economies*,
DOI 10.1007/978-3-319-46439-8

The United Kingdom Standards for Professional Engineering Competence (UK-SPEC)

Graduates from accredited programmes must achieve the following outcomes irrespective of the discipline they enrolled in:

- Knowledge and understanding: they must be able to demonstrate their knowledge and understanding of essential facts, concepts, theories and principles of their engineering discipline, and its underpinning science and mathematics. They must have an appreciation of the wider multidisciplinary engineering context and its underlying principles. They must appreciate the social, environmental, ethical, economic, and commercial considerations affecting the exercise of their engineering judgement.
- Intellectual abilities: they must be able to apply appropriate quantitative science and engineering tools to the analysis of problems. They must be able to demonstrate creative and innovative ability in the synthesis of solutions and in formulating designs. They must be able to comprehend the broad picture and thus work with an appropriate level of detail.
- Practical skills: they must possess practical engineering skills acquired through, for example, work carried out in laboratories and workshops; in industry through supervised work experience; in individual and group project work; in design work; and in the development and use of computer software in design, analysis, and control. Evidence of group work and of participation in a major project is expected. However, individual professional bodies may require particular approaches to this requirement.
- General transferable skills: they must have developed transferable skills that will be of value in a wide range of situations. These are exemplified by the Qualifications and Curriculum Authority Higher Level Key Skills and include problem-solving, communication, and working with others, as well as the effective use of general IT facilities and information retrieval skills. They also include planning self-learning and improving performance, as the foundation for lifelong learning/CPD (Standard and Competence n.d.).

European Union Attributes (Europe)

Generic employability skills proposed by the European Round Table of Industrialists (ERT) are as follows:

- Mastery of one's native language: including the basics of spelling and sentence structure.
- Critical thinking: ability to think through a problem or situation, distinguishing between facts and prejudices.

- Understanding of the basics of math and science: particularly to cope with new technology.
- Learning techniques: ability to pick up new skills and adapting to new situations.
- Team spirit: ability to work in a group.
- Personal discipline: sense of responsibility.
- Decision-making: sense of commitment and willingness to take risks.
- Initiative: sense of curiosity and creativity.
- Professionalism: sense of achieving excellence and gaining competitive edge.
- Civic mindedness: sense of service to the community (Zaharim et al. n.d.).

Engineers Australia (EA)

The Engineers Australia Accreditation Board (2005) has identified similar generic attributes that are as follows:

- Ability to apply knowledge of basic science and engineering fundamentals;
- Ability to communicate effectively, not only with engineers but also with the community at large;
- In-depth technical competence in at least one engineering discipline;
- Ability to undertake problem identification, formulation, and solution;
- Ability to utilize a systems approach to design and operational performance;
- Ability to function effectively as an individual and in multidisciplinary and multicultural teams, with the capacity to be a leader or manager as well as an effective team member;
- Understanding of social, cultural, global, and environmental responsibilities of the professional engineers and the need of sustainable development;
- Understanding of the principles of sustainable design and development;
- Understanding of professional and ethical responsibilities and commitment to them; and
- Expectation of the need to undertake lifelong learning and capacity to do so (Goel 2006).

The Japan Accreditation Board for Engineering Education (JABEE)

The following are the learning outcomes mandated by the board:

- An ability of multidimensional thinking with knowledge from global perspective.
- An ability of understanding of effects and impact of professional activities on society and nature, and of professionals' social responsibility.

- Knowledge of and ability to apply mathematics and natural sciences.
- Knowledge of the related professional fields and the ability to apply.
- Design ability to respond to requirements of the society by utilizing various sciences, technologies, and information.
- Communication skills including logical writing, presentation, and debating.
- An ability of independent and lifelong learning.
- An ability to manage and accomplish tasks systematically under given constraints.
- An ability to work in a team (Criteria et al. 2012).

Engineering Accreditation Councils of Malaysia

Engineering Accreditation Council requires that their accredited programme produce graduates with the following attributes:

- Ability to acquire and apply knowledge of science and engineering fundamentals;
- Ability to communicate effectively, not only with engineers but also with the community at large;
- In-depth technical competence in a specific engineering discipline;
- Ability to undertake problem identification, formulation, and solution;
- Ability to utilize a systems approach to design and evaluate operational performance;
- Understanding of the principles of sustainable design and development;
- Understanding of professional and ethical responsibilities and commitment to them;
- Ability to function effectively as an individual and in a group with the capacity to be a leader or manager as well as an effective team member;
- Understanding of the social, cultural, global, and environmental responsibilities of a professional engineer, and the need for sustainable development; and
- Expectation of the need to undertake lifelong learning and possessing/acquiring the capacity to do so (Programme and Manual 2003).

The Accreditation Criteria of Engineering Programmes at Institution of Engineers, Singapore (IES)

The programme should ensure that graduates are able to:

- Apply knowledge of mathematics, science, and engineering;
- Design and conduct experiments, analyse, interpret data, and synthesize valid conclusions;

- Design a system, component, or process, and synthesize solutions to achieve desired needs;
- Identify, formulate, research through relevant literature review, and solve engineering problems reaching substantiated conclusions;
- Use the techniques, skills, and modern engineering tools necessary for engineering practice with appropriate considerations for public health and safety, cultural, societal, and environmental constraints;
- Communicate effectively;
- Recognize the need for and have the ability to engage in lifelong learning;
- Understand the impact of engineering solutions in a societal context and to be able to respond effectively to the needs for sustainable development;
- Function effectively within multidisciplinary teams and understand the fundamental precepts of effective project management; and
- Understand professional, ethical, and moral responsibility (Engineering Accreditation Accreditation 2004).

Appendix B
Generic Definitions for Set of Mutual 22 Skills

Dimension I: Core Engineering Knowledge and Practice

Math, Physics, Science Fundamentals (MPSF) has been referred in the literature using terms such as "numeric modelling" (Male 2010a, b) and "literacy and numeracy skills" (Knight 2012); as well as "science and mathematics fundamentals" (Rajala 2012; Hundley 2012; Mishra 2010; Chan et al. 2009) which is the most commonly encountered. MPSF was emphasized in about 41 % of the GEND literature as compared to 24 % in the ENGD. A proposed generic definition of (MPSF) is as follows: "demonstration of essential mathematics skills and use, quantify and manipulate numbers and their application accurately in any context".
Disciplinary Knowledge (Depth) has been encountered in the literature under different terms such as "Engineering Science" (Duderstad); "Engineering Fundamentals" (kerr 2010); and "Theoretical Knowledge" (Michigan 2009). Depth knowledge has been emphasized in about 41 % of the GEND literature and 43 % of the ENGD literature. A proposed generic definition is as follows: "demonstrate in-depth technical knowledge in the engineering field of specialty and know-how in specific engineering discipline and apply this knowledge effectively through practice—in balance with non-technical skills".
Interdisciplinary Knowledge (Breadth) was referred to in the literature using different terms such as "interdisciplinary knowledge" (Shuman et al. 2000a, b) and "interdisciplinary skills" (Knight 2012; NSB 2007). Interestingly, interdisciplinary knowledge was emphasized in 16 % of the ENGD literature but with not emphasis in the GEND literature. A proposed generic definition is as follows: "develop and build a multidisciplinary knowledge from other engineering fields and a comprehensive interdisciplinary knowledge from a specific filed through a scholarly approach that incorporating theory and practice".
Multidisciplinary Knowledge (MDK) was referred to as "multidisciplinary" in most of the literature (NAE 2004; Hundley 2012; Rajala 2012; Chang et al. 2009; Continental 2006) and sometimes as "other fields" (Palmer et al. 2011). MDK was emphasized in 6 % of the GEND literature and 27 % of the ENGD literature. A proposed generic definition is as follows: "develop and build a multidisciplinary knowledge from other engineering fields and a comprehensive interdisciplinary

M. Abdulwahed and M.O. Hasna, *Engineering and Technology Talent for Innovation and Knowledge-Based Economies*,
DOI 10.1007/978-3-319-46439-8

knowledge from a specific filed through a scholarly approach that incorporating theory and practice".

Practical Skills (PrS) term has been explicitly mentioned in a wide variety of literature (Male 2010a; Markes 2006), and also, the term has been coined with relevant terms and phrases in other literature, such as "COMPETENT IN APPLICATION AND PRACTICE" (Zaharim et al. 2009), "EXHIBIT PRACTICAL INGENUITY"(Mena et al. 2012), and "PRACTICAL KNOWLEDGE AND WISDOM"(Michigan 2009). "PRS" was emphasized in 39 % of the ENGD literature and 32 % of the GEND literature. A proposed generic definition of "PRS" extracted from the different ENGD and GEND references is as follows: "*apply knowledge and skills, required for effective professional practices, into real-world settings*".

Information and Computer Technology Skills (ICTS) term has been coined with relevant terms and phrases in other literature, such as "ability to apply modern ICT" (Continental 2006); "knowledge of contemporary issues (ICT)" (Zaharim et al. 2009); "information literacy (ICT)" (ACER 2002; Bowman 2010; Edinburgh 2012; Finegold and Notabartolo 2010); "general IT skills" (Martin et al. 2008; PyeTait 2011; UNESCO 2012); and "workplace information and communications technology" (WDA 2009; Zaharim et al. 2009). "ICTS" was emphasized in 24 % of the ENGD literature and 76 % of the GEND literature. A proposed generic definition of "ICTs" extracted from the different ENGD and GEND references is as follows: "use information and communication technology applications interactively in locating, managing, modelling, analysing, evaluating, and processing information from multiple sources to effectively accomplish a given task".

Dimension II: Cognition and Thinking

Problem-Solving Skills (PSS) term has been explicitly mentioned in a wide variety of literature (Allan and Chisholm 2009; CBI 2007, 2009, 2012; Llanes 2008; NESP 2010; NHEAP 2007; NSB 2007; Shuman et al. 2000a, b), and also, the term has been coined with relevant terms and phrases in other literature, such as "ability to frame problems, putting them in a socio-technical and operational context" (NAE 2004, 2005; Sunthonkanokpong 2011) and "the ability to analyse and solve complex problems"(HRA 2013). "PSS" was emphasized in 65 % of the ENGD literature and 91 % of the GEND literature. A proposed generic definition of "PSS" extracted from the different ENGD and GEND references is as follows: "*use knowledge systematically to identify, analyse, formulate, solve, and evaluate complex and multidisciplinary problems—by applying cognitive skills (logical, critical, and creative thinking)*".

Lifelong learning (LLL) term has been explicitly mentioned in a wide variety of literature (NRC 2004, 2005; Kerr 2010; Chan et al. 2009; Mishra 2010; Sunthonkanokpong 2011; Continental 2006; NSB 2007; IEA 2009; Llanes 2008) and also, the term has been coined with relevant terms and phrases in other literature, such as "*curiosity and desire to learn for life*" (Chang et al. 2009; Rajala

2012) and "*continuous learners*" (Jamieson et al. 2009). "LLL" was emphasized in 57 % of the ENGD literature and 53 % of the GEND literature. A proposed generic definition of "LLL" extracted from the different ENGD and GEND references is as follows: "*continuously acquire new knowledge and skills for self and professional development at all levels (life and career)*".

Decision-Making Skills (DMS) term has been explicitly mentioned in a wide variety of literature (Jamieson et al. 2009; Michigan 2009; Finegold and Notabartolo 2010; SHAH 2009; Bowman 2010; UNESCO 2012; Zaharim et al. 2009) and also, the term has been coined with relevant terms and phrases in other literature, such as "*judgment and decision making*" (Tippins and Hilton 2010) and "*applies personal and professional judgment in effectively making decisions and managing risks*" (Hundley et al. 2012; IFEES n.d.). "DMS" was emphasized in 14 % of the ENGD literature and 32 % of the GEND literature. A proposed generic definition of "DMS" extracted from the different ENGD and GEND references is as follows: "*apply personal and professional judgment, take-risk and initiative, in effectively making strategic decision and managing risks, from a range of alternatives, based-on available information in response to ambiguous and complex situation*".

System Thinking approach (STA) term has been explicitly mentioned in a wide variety of literature (Michigan 2009; Palmer et al. 2011; NSB 2007; NRC 2011) and also, the term has been coined with relevant terms and phrases in other literature, such as "systems skills (analysis/ evaluation/Operations analysis)"(Tippins and Hilton 2010); "engineering system approach" (Zaharim et al. 2009); and "systems approach to engineering" (Rajala 2012; Chang et al. 2009). "STA" was emphasized in 30 % of the ENGD literature and 12 % of the GEND literature. A proposed generic definition of "STA" extracted from the different ENGD and GEND references is as follows: "*think systematically in holistic approach by understanding how an entire system works and the relationship among systems' elements and other systems, and viewing a problem as a part of an overall system*".

Critical Thinking (ClT) term has been explicitly mentioned in a wide variety of literature (Allan and Chisholm 2009; Danielson 2011; Michigan 2009; Shuman et al. 2000a, b; Tippins and Hilton 2010), and also, the term has been coined with relevant terms and phrases in other literature, such as "ability to think critically (Chang et al. 2009; IFEES n.d.; Rajala 2012) and "the ability to apply a systematic and critical assessment of complex problems and issues" (Hounsell 2011). "ClT" was emphasized in 32 % of the ENGD literature and 44 % of the GEND literature. A proposed generic definition of "ClT" extracted from the different ENGD and GEND references is as follows: "*use critical, conceptual, reflective, and rational thinking in drawing and evidence-based assessing systematic conclusions and finding underlying relationships for solutions*".

Innovation Skills (InS) term has been explicitly mentioned in a wide variety of literature (Mishra 2010; Bowman 2010; Chang et al. 2009; Finegold and Notabartolo 2010; HRA 2013; P21 2008; Palmer et al. 2011; Terenzini and Lattuca n.d.), and also, the term has been coined with relevant terms and phrases in other literature, such as "innovative solutions" (Mena et al. 2012); "an ability to innovate" (NSB 2007); "*technological innovations*" (Kerr 2010); and "*ability to develop new or innovative ideas*" (Patil and Codner 2008). "InS" was emphasized in 32 % of the ENGD literature and 18 % of the GEND literature. A proposed generic definition of "InS" extracted from the different ENGD and GEND references is as follows: "*add values through introducing new novel ideas, methods, directions, opportunities, and solutions that meet new requirements, through more effective products, processes, services, and technologies that are readily available to stakeholders*".

System Design Skills (SDS) term has been explicitly mentioned in the literature (Mishra 2010)."SDS" term has been coined with relevant terms and phrases in other literature, such as "*design skills*" (Jamieson et al. 2009; Knight 2012; Chan et al. 2010); "design—product creation"; and "engineering design" "*Technology design*" (Tippins and Hilton 2010). "SDS" was emphasized in 41 % of the ENGD literature and 6 % of the GEND literature. A proposed generic definition of "SDS" extracted from the different ENGD and GEND references is as follows: "*design products to meet specified needs and develop optimal solutions for complex problems*".

Dimension III: Professional and Interpersonal

Professionalism (PsS) term has been explicitly mentioned in a wide variety of literature (NRC 2004, 2005; UNESCO 2012), and also, the term has been coined with relevant terms and phrases in other literature, such as "*professional behaviours and accountability*" (Edinburgh 2012) and "*maintains a high-level of professional competence*" (IFEES n.d.). "PsS" was emphasized in 24 % of the ENGD literature and 21 % of the GEND literature. A proposed generic definition of "PsS" extracted from the different ENGD and GEND references is as follows: "*maintain high-level of social and professional behavior, demonstrate accountability, accept responsibility, and achieve excellence in work and everyday life*".

Ethics and Responsibility (E&R) term has been coined with relevant terms and phrases in other literature, such as "*high ethical standards*" (IEA 2009; NAE 2004, 2005; Rajala 2012); "*ethical issues in engineering practice*" (Palmer et al. 2011; Terenzini and Lattuca n.d.); and identify ethical implications (Mena et al. 2012). "E&R" was emphasized in 57 % of the ENGD literature and 26 % of the GEND literature. A proposed generic definition of "E&R" extracted from the different ENGD and GEND references is as follows: "*understand ethical norms and demonstrate ethical reasoning, behaviour, and integrity of social responsibility and service in any context (work, community, country, etc.…)*".

Communication Skills (CS) term has been explicitly mentioned in a wide variety of literature (Chang et al. 2009; IEA 2009; Jamieson et al. 2009; Kerr 2010; Knight 2012; Mena et al. 2012; Rabl and Hillmer 2012; Shuman et al. 2000a, b). "CS" has been coined with relevant terms and phrases in other literature, such as "*effective communication*" (Nguyen 1998; Rajala 2012) and "*interpersonal skills with colleagues and clients/Oral communication skills/Written communication skills*" (Patil and Codner 2008). "CS" was emphasized in 92 % of the ENGD literature and 97 % of the GEND literature. A proposed generic definition of "CS" extracted from the different ENGD and GEND references is as follows: "*communicate effectively, in various ways (verbal, oral, written, ...), with all stakeholders (Public, Engineers, ...) across all boundaries (cultural, language, ...)*".

Team Work Skills (TWS) term has been coined with different relevant terms and phrases in the literature, such as "*working effectively in teams*" (Palmer et al. 2011; Terenzini and Lattuca n.d.); "*capacity for co-operation and teamwork*" (Patil and Codner 2008); "*teams with people from different countries/Globally distributed teams/teams with people from other disciplines*" (Mena et al. 2012); and "*individual/teamwork skills in multidisciplinary and multicultural team*" (Mishra 2010). "TWS" was emphasized in 73 % of the ENGD literature and 88 % of the GEND literature. A proposed generic definition of "TWS" extracted from the different ENGD and GEND references is as follows: "*the ability to work and cooperate efficiently in a diverse, multicultural, and interdisciplinary team*".

Foreign Language Skills (FLS) term has been explicitly mentioned in a wide variety of literature (UKCES 2009; P21 2008; Chang et al. 2009; SHAH 2009; CBI 2007), and also, the term have been coined with relevant terms and phrases in other literature, such as "*ability to write and speak in a foreign language*" (REFLEX 2007); "*should be bilingual*" (Nguyen 1998); and "*possesses fluency in at least two languages*" (IFEES n.d.). "FLS" was emphasized in 22 % of the ENGD literature and 38 % of the GEND literature. A proposed generic definition of "FLS" extracted from the different ENGD and GEND references is as follows: "*demonstrate language proficiency, in English and the native language, at both level, professional and technical*".

Dimension IV: Business and Management

Business and Management skills (PMS) term has been explicitly mentioned in a wide variety of literature (Markes 2006; Michigan 2009; NAE 2004, 2005; NHEAP 2007; Rabl and Hillmer 2012; Sunthonkanokpong 2011) and also, the term has been coined with relevant terms and phrases in other literature, such as "*project management and finance*" (IEA 2009) and "*an ability to understand the business context of engineering*" (NSB 2007). "PMS" was emphasized in 76 % of the ENGD literature and 71 % of the GEND literature. A proposed generic definition of "PMS" extracted from the different ENGD and GEND references is as follows: "*awareness of business literacy and practice in the context of engineering; manage organization and resources efficiently; and develop processes, plan projects, and take-risk*".

Leadership Skills (LsS) term has been explicitly mentioned in a wide variety of literature (Chang et al. 2009; Finegold and Notabartolo 2010; Fung et al. 2006; Hounsell 2011; Jamieson et al. 2009; Markes 2006; Nguyen et al. 2005; P21 2008; Palmer et al. 2011; Rajala 2012; SHAH 2009), and also, the term has been coined with relevant terms and phrases in other literature, such as "*ability to assert your authority*" (REFLEX 2007) and "*ability to advocate and influence*". "LsS" was emphasized in 51 % of the ENGD literature and 35 % of the GEND literature. A proposed generic definition of "LsS" extracted from the different ENGD and GEND references is as follows: "*engage, interact, lead, and influence people effectively toward accomplishing a common or shared goal and contributing positively to diverse communities*".

Entrepreneurship Skills (EpS) term has been explicitly mentioned in a wide variety of literature (Danielson 2011; Chang et al. 2009; Rajala 2012; Allan and Chisholm 2009; Edinburgh 2012; Fung et al. 2006; UNESCO 2012; NSB 2007), and also, the term has been coined with relevant terms and phrases in other literature, such as "*entrepreneurial thinking*"; "*an entrepreneurial mind*" (Nguyen et al. 2005) and "*Entrepreneurial and intrapreneurial*". "EpS" was emphasized in 22 % of the ENGD literature and 18 % of the GEND literature. A proposed generic definition of "EpS" extracted from the different ENGD and GEND references is as follows: "*identify, capture, and develop an opportunity to create real business and social value through converting invention into successful innovation*".

Appendix C
Research Methodologies in the Literature

See Table C.1.

Table C.1 Summary of Research Methodologies

Reference	Methodology components	Data analysis method (qualitative/quantitative)	Tools availability
USA (Palmer et al. 2011)	Literature review, surveys	Mixed	Partially
USA (Jamieson et al. 2009)	Workshops, discussions, surveys, focus groups	Mixed	No
USA (Mena et al. 2012)	Survey, telephone interviews	Mixed	No
Canada (Chan and Fishbein 2009)	Focus groups	Qualitative	No
US (Pritchard 2013)	Survey, interview, literature review	Mixed	No
AUS (Patil and Codner 2008)	Survey	Quantitative	Partially
USA (Hundley et al. 2012, 2013)	Literature search, survey	Quantitative	Partially
USA (Parkinson 2009)	Survey	Quantitative	No
USA (Warrington and Tech 2010)	Discussions, survey	Mixed	No
USA (Knight 2012)	Literature review, focus groups, survey, interviews	Mixed	Partially
USA (NAE 2004, 2005)	A summit for vision articulation	Mixed	No
UK (RAE 2007)	Interviews, focus groups, surveys	Mixed	No
UK (Spinks et al. 2006)	Interviews, focus groups, surveys	Mixed	Yes
USA (NSF 2007)	Workshops, presentation and discussion, meetings	Qualitative	No

(continued)

M. Abdulwahed and M.O. Hasna, *Engineering and Technology Talent for Innovation and Knowledge-Based Economies*,
DOI 10.1007/978-3-319-46439-8

Table C.1 (continued)

Reference	Methodology components	Data analysis method (qualitative/quantitative)	Tools availability
UK (Bourn and Neal 2008)	Workshops, discussion	Qualitative	No
UK (Dudman and Wearne 2003)	Survey	Quantitative	Yes
USA (ASME 2008)	A summit for vision articulation	Qualitative	No
USA (Michigan Engineering 2020 2009)	Survey, focus groups, discussions	Mixed	No
USA (ASCE 2007)	A summit for vision articulation (involving surveys and discussion)	Mixed	Yes
AUS (Graduate Employability Skills 2007)	Interviews, discussion papers	Qualitative	No
AUS (King 2008)	Discussion, focus groups consultation	Qualitative	Yes
AUS (Employability Skills Framework 2012)	Consultations	Qualitative	No
US (Rajala 2012)	Survey	Quantitative	Partially
Japan (Nguyen et al. 2005)	Survey	Mixed	No
AUS (NCVER 2003)	Literature review (multidisciplinary dimensions)	Qualitative	No
AUS (Male 2009)	Survey	Quantitative	Partially
Malaysia (Saad et al. 2013)	Survey	Quantitative	Partially
US (ASME n.d.)	Survey	Quantitative	No
Egypt (Osman 2011)	Interview, survey	Mixed	Yes
US (Duderstadt 2008b)	Roadmap	Qualitative	No
London (Martin et al. 2008)	Survey	Mixed	Yes
US (POMALES-GARCÍA and LIU 2007)	Focus groups	Qualitative	Yes
Asia (UNESCO, Indonesia: Survey 2012)	Interviews	Mixed	No
	Malaysia: focus groups, resources review	Qualitative	No
AUS (Rabl 2012)	Survey	Quantitative	No

Appendix D
Targeted Groups Survey Design

Faculty Members Survey Design: **Section 1**: **Demographics** asking some data about their nationality, genders, detailed educational background, undergraduate engineering programme teaching and to which group(s), number of years of experience detailing those that are outside Qatar, and if there was any previous industrial experience or not. **Section 2**: **Competencies and skills of engineering graduates in Qatar**: asking about twenty-first-century skills importance currently and by 2030 as well as their satisfaction with such skills currently asking them to rank the top 5 skills from them asking about their opinions regarding Qatar industry demand of these skills and if it will change by 2022 and by 2030, in addition to a comparison between students groups relevant to genders, nationality, residency, and universities and finally their agreement level with some potential barriers that they believe will hinder students preparedness with required skills. **Section 3: Undergraduate engineering Education in Qatar**: dealing with their agreement level in how much engineering education institutions will help prepare graduates for their work places. **Section 4: Industry–academia collaboration**: asking if there is any collaboration between academia and industry and to sate preferable means for them they aim their universities do or already did and to state if there is any barriers hindering the collaboration with industry and state its kind.

QU Students Survey Design: **Section 1**: **Demographics**: asking some data about their nationality, genders, GPA, completed credit hours, and finally their engineering major. **Section 2**: **Competencies and skills of engineering graduates in Qatar**: asking about twenty-first-century skills importance currently and by 2030 as well as their satisfaction with such skills currently asking them to rank the top 5 skills from them asking about their opinions regarding Qatar industry demand of these skills and if it will change by 2022 and by 2030, asking them state if they have any courses or experiences before that may helped in gaining or improving any of these skills. **Section 3**: **Undergraduate engineering Education in Qatar**: dealing with their agreement level in some teaching styles appropriateness in preparing and qualifying students for the engineering practice and asking them to state from their point of view their level of agreement with students competencies, knowledge, language proficiency, and practical experience comparing QU students with those of educational city and overseas ones. **Section 4**: **Training, Practical experience, and employment**: asking them if they have been involved in any

M. Abdulwahed and M.O. Hasna, *Engineering and Technology Talent for Innovation and Knowledge-Based Economies*,
DOI 10.1007/978-3-319-46439-8

internship or training in their undergraduate studies sating the beneficial skills behind such training course, and asking them about their career plan after graduation, the easiness to get employed and if there were any barriers that they faced or may face.

Professional engineers/Alumni Survey design: **Section 1: Demographics** asking students some data about their nationality, genders, educational background, employment status, years of experience, employment role and function, first employment timing, and average time they took to be expert in their current work. **Section 2**: **Competencies and skills of engineering graduates in Qatar**: asking about twenty-first-century skills importance currently and by 2030 as well as their satisfaction with such skills currently asking them to rank the top 5 skills from them asking about their opinions regarding Qatar industry demand of these skills and if it will change by 2022 and by 2030, and asking them state if they have any courses or experiences before that may helped in gaining or improving any of these skills. **Section 3: Undergraduate engineering Education in Qatar**: dealing with their agreement level in some teaching styles appropriateness in preparing and qualifying students for the engineering practice and asking them to state from their point of view their level of agreement with students competencies, knowledge, language proficiency, and practical experience comparing QU students with those of educational city and overseas ones and finally letting them to get back and have the choice and see whether they will choose same study programme and college or not. **Section 4: Training, Practical experience and employment:** asking them if they have been involved in any internship or training in their undergraduate studies sating the beneficial skills behind such training course, and asking them about the easiness to get employed and if there were any barriers that they faced.

Industry Survey Design: **Section 1**: **Demographics**: asking students some data about their nationalities, genders, and work position. **Section 2**: **Competencies and skills of engineering graduates in Qatar**: asking about twenty-first-century skills importance currently and by 2030 as well as their satisfaction with such skills currently asking them to rank the top 5 skills from them asking about their opinions regarding Qatar industry demand of these skills and if it will change by 2022 and by 2030, in addition to a comparison between students groups relevant to genders, nationality, residency, and universities and finally their agreement level with some potential barriers that they believe will hinder students preparedness with required skills. **Section 3**: **Industry–academia collaboration**: asking if there is any collaboration between academia and industry and to sate preferable means for them, they aim the universities will do or already did and to state if there is any barriers hindering the collaboration with industry and state its kind.

Appendix E
Engineering Skills and Competency Analysis Between Junior Engineers and Senior Students

See Tables E.1, E.2, E.3 and E.4.

Table E.1 The statistics of the students' evaluation of twenty-second-century skills importance currently and by 2030 their satisfaction level between alumni and senior students perceptions "Dimension I"

Dimension I: Core engineering knowledge and practice						
Skills	Mean		Sample number (N)	P value (Mann–Whitney U test)	P value (independent sample t test)	Statistical significance
Importance currently						
Disciplinary engineering fundamentals (depth)	Alumni	4.32	28	0.401	0.260	No
	Senior students	4.09	287			
Interdisciplinary engineering fundamentals (breadth)	Alumni	4.00	28	0.529	0.931	No
	Senior students	4.02	283			
Math, physics, and science fundamentals	Alumni	4.15	27	0.831	0.598	No
	Senior students	4.03	281			
Practical experience	Alumni	4.67	27	0.149	0.077	No
	Senior students	4.29	276			
ICT experience	Alumni	4.41	27	0.270	0.133	No
	Senior students	4.07	286			
Multidisciplinary knowledge	Alumni	3.71	28	0.104	0.221	No
	Senior students	3.98	281			

(continued)

M. Abdulwahed and M.O. Hasna, *Engineering and Technology Talent for Innovation and Knowledge-Based Economies*,
DOI 10.1007/978-3-319-46439-8

Table E.1 (continued)

Dimension I: Core engineering knowledge and practice						
Skills	Mean		Sample number (N)	P value (Mann–Whitney U test)	P value (independent sample t test)	Statistical significance
Importance 2030						
Disciplinary engineering fundamentals (depth)	Alumni	4.57	28	0.189	0.171	No
	Senior students	4.30	288			
Interdisciplinary engineering fundamentals (breadth)	Alumni	4.50	28	0.169	0.107	No
	Senior students	4.16	287			
Math, physics, and science fundamentals	Alumni	4.30	27	0.371	0.256	No
	Senior students	4.03	288			
Practical experience	Alumni	4.81	27	0.132	0.059	No
	Senior students	4.43	282			
ICT experience	Alumni	4.63	27	0.095	0.082	No
	Senior students	4.24	288			
Multidisciplinary knowledge	Alumni	4.00	28	0.445	0.678	No
	Senior students	4.09	281			
Satisfaction						
Disciplinary engineering fundamentals (depth)	Alumni	3.74	23	0.360	0.353	No
	Senior students	3.53	287			
Interdisciplinary engineering fundamentals (breadth)	Alumni	3.50	22	0.941	0.849	No
	Senior students	3.46	283			
Math, physics, and science fundamentals	Alumni	3.91	22	0.382	0.285	No
	Senior students	3.66	288			
Practical experience	Alumni	3.38	21	0.992	0.938	No
	Senior students	3.36	284			
ICT experience	Alumni	3.86	22	0.072	0.099	No
	Senior students	3.47	287			
Multidisciplinary knowledge	Alumni	3.29	21	0.801	0.652	No
	Senior students	3.40	286			

Table E.2 The statistics of the students' evaluation of twenty-second-century skills importance currently and by 2030 their satisfaction level between alumni and senior students perceptions "Dimension II"

Dimension II: Cognition and thinking						
Skills	Mean		Sample number (N)	P value (Mann–Whitney U test)	P value (independent sample t test)	Statistical significance
Importance currently						
Lifelong learning	Alumni	4.50	28	0.213	0.189	No
	Senior students	4.23	278			
Problem-solving	Alumni	4.54	28	0.822	0.355	No
	Senior students	4.35	283			
Decision-making	Alumni	4.32	28	0.648	0.528	No
	Senior students	4.19	288			
Systems thinking	Alumni	4.11	28	0.636	0.788	No
	Senior students	4.16	288			
Critical thinking	Alumni	4.21	28	0.514	0.864	No
	Senior students	4.25	284			
Innovation	Alumni	4.25	28	0.756	0.849	No
	Senior students	4.21	286			
Design	Alumni	4.41	27	0.703	0.498	No
	Senior students	4.26	284			
Importance 2030						
Lifelong learning	Alumni	4.64	28	0.137	0.153	No
	Senior students	4.34	280			
Problem-solving	Alumni	4.61	28	0.407	0.339	No
	Senior students	4.41	287			
Decision-making	Alumni	4.54	28	0.501	0.437	No
	Senior students	4.38	289			
Systems thinking	Alumni	4.29	28	0.731	0.942	No
	Senior students	4.30	289			
Critical thinking	Alumni	4.39	28	0.932	0.758	No
	Senior students	4.33	290			

(continued)

Table E.2 (continued)

Dimension II: Cognition and thinking						
Skills	Mean		Sample number (N)	P value (Mann–Whitney U test)	P value (independent sample t test)	Statistical significance
Innovation	Alumni	4.57	28	0.789	0.343	No
	Senior students	4.38	289			
Design	Alumni	4.63	27	0.609	0.250	No
	Senior students	4.37	288			
Satisfaction						
Lifelong learning	Alumni	3.82	22	0.271	0.383	No
	Senior students	3.61	287			
Problem-solving	Alumni	3.64	22	0.718	0.771	No
	Senior students	3.71	287			
Decision-making	Alumni	3.41	22	0.631	0.609	No
	Senior students	3.53	288			
Systems thinking	Alumni	3.68	22	0.901	0.988	No
	Senior students	3.68	286			
Critical thinking	Alumni	3.55	22	0.971	0.878	No
	Senior students	3.58	287			
Innovation	Alumni	3.59	22	0.399	0.517	No
	Senior students	3.43	285			
Design	Alumni	3.73	22	0.795	0.983	No
	Senior students	3.72	284			

Table E.3 The statistics of the students' evaluation of twenty-second-century skills importance currently and by 2030 their satisfaction level between alumni and senior students perceptions "Dimension III"

Dimension III: Professional and interpersonal						
Skills	Mean		Sample number (N)	P value (Mann–Whitney U test)	P value (independent sample t test)	Statistical significance
Importance currently						
Professionalism	Alumni	4.43	28	0.423	0.582	No
	Senior students	4.26	285			
Ethics	Alumni	4.57	28	0.302	0.399	No
	Senior students	4.35	284			
Adaptability	Alumni	4.29	28	0.542	0.693	No
	Senior students	4.17	278			
Communication	Alumni	4.54	28	0.289	0.332	No
	Senior students	4.31	281			
Teamwork	Alumni	4.61	28	0.173	0.172	No
	Senior students	4.30	285			
Foreign language(s)	Alumni	4.56	27	0.154	0.285	No
	Senior students	4.26	286			
Importance 2030						
Professionalism	Alumni	4.59	27	0.445	0.331	No
	Senior students	4.40	285			
Ethics	Alumni	4.68	28	0.452	0.237	No
	Senior students	4.43	288			
Adaptability	Alumni	4.39	28	0.694	0.535	No
	Senior students	4.26	280			
Communication	Alumni	4.71	28	0.072	0.149	No
	Senior students	4.40	284			
Teamwork	Alumni	4.79	28	**0.027**	0.076	Yes/No
	Senior students	4.39	289			
Foreign language(s)	Alumni	4.56	27	0.440	0.243	No
		4.30	289			

(continued)

Table E.3 (continued)

Dimension III: Professional and interpersonal						
Skills	Mean		Sample number (N)	P value (Mann–Whitney U test)	P value (independent sample t test)	Statistical significance
	Senior					
Satisfaction						
Professionalism	Alumni	3.45	22	0.344	0.392	No
	Senior students	3.66	286			
Ethics	Alumni	3.73	22	0.928	0.973	No
	Senior students	3.72	284			
Adaptability	Alumni	3.59	22	0.547	0.760	No
	Senior students	3.66	277			
Communication	Alumni	3.86	22	0.933	0.989	No
	Senior students	3.87	286			
Teamwork	Alumni	4.18	22	0.101	0.144	No
	Senior students	3.81	289			
Foreign language(s)	Alumni	4.05	22	0.214	0.196	No
	Senior students	3.70	289			

Table E.4 The statistics of the students' evaluation of twenty-second-century skills importance currently and by 2030 their satisfaction level between alumni and senior students perceptions "Dimension IV"

Dimension IV: Business management						
Skills	Mean		Sample number (N)	P value (Mann–Whitney U test)	P value (independent sample t test)	Statistical significance
Importance currently						
Management	Alumni	4.14	28	0.564	0.981	No
	Senior students	4.15	284			
Leadership	Alumni	4.32	28	0.648	0.525	No
	Senior students	4.19	283			
Entrepreneurship	Alumni	4.07	27	0.409	0.705	No
	Senior students	4.16	279			
Importance 2030						
Management	Alumni	4.56	27	0.857	0.348	No
	Senior students	4.36	288			
Leadership	Alumni	4.67	27	0.202	0.102	No
	Senior students	4.32	286			
Entrepreneurship	Alumni	4.38	26	0.988	0.700	No
	Senior students	4.30	285			
Satisfaction						
Management	Alumni	3.23	22	0.344	0.330	No
	Senior students	3.46	280			
Leadership	Alumni	3.68	22	0.754	0.756	No
	Senior students	3.60	281			
Entrepreneurship	Alumni	3.27	22	0.569	0.617	No
	Senior students	3.40	277			

Appendix F
Engineering Skills and Competencies Between Faculty Members and Industry

See Tables F.1, F.2 and F.3.

Table F.1 The statistics of the students' evaluation of twenty-second-century skills importance currently and by 2030 their satisfaction level between faculty members and industrial sector perceptions "Dimension I"

Dimension I: Core engineering knowledge and practice						
Skills	Mean		Sample number (N)	P value (Mann–Whitney U test)	P value (independent sample t test)	Statistical significance
Importance currently						
Disciplinary engineering fundamentals (depth)	Faculty	4.44	41	0.982	0.979	No
	Industry	4.44	61			
Interdisciplinary engineering fundamentals (breadth)	Faculty	4.22	41	0.300	0.315	No
	Industry	4.05	61			
Math, physics, and science fundamentals	Faculty	4.22	41	0.875	0.695	No
	Industry	4.15	59			
Practical experience	Faculty	4.22	41	**0.033**	**0.063**	**Yes/No**
	Industry	4.52	60			
ICT experience	Faculty	4.41	41	0.756	0.697	No
	Industry	4.36	59			
Multidisciplinary knowledge	Faculty	3.78	40	0.827	0.852	No
	Industry	3.74	61			
Importance 2030						
Disciplinary engineering	Faculty	4.59	41	0.759	0.809	No
	Industry	4.62	60			

(continued)

M. Abdulwahed and M.O. Hasna, *Engineering and Technology Talent for Innovation and Knowledge-Based Economies*,
DOI 10.1007/978-3-319-46439-8

Table F.1 (continued)

Dimension I: Core engineering knowledge and practice						
Skills	Mean		Sample number (N)	P value (Mann–Whitney U test)	P value (independent sample t test)	Statistical significance
fundamentals (depth)						
Interdisciplinary engineering fundamentals (breadth)	Faculty	4.54	41	0.937	0.698	No
	Industry	4.48	60			
Math, physics, and science fundamentals	Faculty	4.39	41	0.699	0.490	No
	Industry	4.28	58			
Practical experience	Faculty	4.61	41	0.057	0.107	No
	Industry	4.78	59			
ICT experience	Faculty	4.66	41	0.597	0.979	No
	Industry	4.66	58			
Multidisciplinary knowledge	Faculty	4.15	41	0.938	0.803	No
	Industry	4.10	60			
Satisfaction						
Disciplinary engineering fundamentals (depth)	Faculty	3.68	40	0.599	0.905	No
	Industry	3.65	55			
Interdisciplinary engineering fundamentals (breadth)	Faculty	3.45	40	0.199	0.248	No
	Industry	3.22	54			
Math, physics, and science fundamentals	Faculty	3.28	40	0.072	0.082	No
	Industry	3.60	53			
Practical experience	Faculty	3.30	40	0.407	0.469	No
	Industry	3.13	53			
ICT experience	Faculty	3.68	40	0.533	0.636	No
	Industry	3.76	54			
Multidisciplinary knowledge	Faculty	3.33	40	0.525	0.393	No
	Industry	3.13	53			

Table F.2 The statistics of the students' evaluation of twenty-second-century skills importance currently and by 2030 their satisfaction level between faculty members and industrial sector perceptions "Dimension II"

Dimension II: Cognition and thinking						
Skills	Mean		Sample number (N)	P value (Mann–Whitney U test)	P value (independent sample t test)	Statistical significance
Importance currently						
Lifelong learning	Faculty	4.20	41	0.053	0.104	No
	Industry	4.46	61			
Problem-solving	Faculty	4.29	41	0.428	0.385	No
	Industry	4.43	61			
Decision-making	Faculty	3.98	41	0.060	0.192	No
	Industry	4.21	62			
Systems thinking	Faculty	3.95	41	0.127	0.231	No
	Industry	4.16	61			
Critical thinking	Faculty	4.22	41	0.616	0.749	No
	Industry	4.28	61			
Innovation	Faculty	4.00	41	0.222	0.167	No
	Industry	4.26	62			
Design	Faculty	4.31	39	0.923	0.655	No
	Industry	4.23	61			
Importance 2030						
Lifelong learning	Faculty	4.63	41	0.189	0.487	No
	Industry	4.72	60			
Problem-solving	Faculty	4.78	41	0.238	0.153	No
	Industry	4.62	60			
Decision-making	Faculty	4.63	41	0.594	0.524	No
	Industry	4.55	62			
Systems thinking	Faculty	4.44	41	0.820	0.932	No
	Industry	4.43	61			
Critical thinking	Faculty	4.68	41	0.525	0.361	No
	Industry	4.56	61			
Innovation	Faculty	4.51	41	0.616	0.422	No
	Industry	4.61	62			
Design	Faculty	4.56	39	0.887	0.986	No
	Industry	4.57	60			
Satisfaction						
Lifelong learning	Faculty	3.44	41	0.897	0.981	No
	Industry	3.44	54			
Problem-solving	Faculty	3.22	41	0.744	0.702	No
	Industry	3.30	53			

(continued)

Table F.2 (continued)

Dimension II: Cognition and thinking						
Skills	Mean		Sample number (N)	P value (Mann–Whitney U test)	P value (independent sample t test)	Statistical significance
Decision-making	Faculty	3.10	41	0.442	0.549	No
	Industry	3.22	55			
Systems thinking	Faculty	3.33	40	0.602	0.607	No
	Industry	3.42	55			
Critical thinking	Faculty	3.20	41	1.000	0.952	No
	Industry	3.18	55			
Innovation	Faculty	3.08	40	0.781	0.878	No
	Industry	3.11	55			
Design	Faculty	3.74	38	0.051	0.053	No
	Industry	3.35	55			

Table F.3 The statistics of the students' evaluation of twenty-second-century skills importance currently and by 2030 their satisfaction level between faculty members and industrial sector perceptions

Dimension III: Professional and interpersonal						
Skills	Mean		Sample number (N)	P value (Mann–Whitney U test)	P value (independent sample t test)	Statistical significance
Importance currently						
Professionalism	Faculty	4.37	41	0.399	0.667	No
	Industry	4.44	62			
Ethics	Faculty	4.41	41	0.317	0.544	No
	Industry	4.52	62			
Adaptability	Faculty	3.93	41	**0.011**	**0.022**	**Yes**
	Industry	4.32	62			
Communication	Faculty	4.44	41	0.164	0.541	No
	Industry	4.53	62			
Teamwork	Faculty	4.34	41	**0.007**	**0.083**	**Yes/No**
	Industry	4.61	62			
Foreign language(s)	Faculty	4.29	41	**0.023**	**0.048**	**Yes**
	Industry	4.57	60			
Importance 2030						
Professionalism	Faculty	4.68	41	0.769	0.961	No
	Industry	4.69	61			
Ethics	Faculty	4.71	41	0.666	0.632	No

(continued)

Table F.3 (continued)

Dimension III: Professional and interpersonal						
Skills	Mean		Sample number (N)	P value (Mann–Whitney U test)	P value (independent sample t test)	Statistical significance
			62			
Adaptability	Faculty	4.49	41	0.564	0.758	No
	Industry	4.53	62			
Communication	Faculty	4.76	41	0.713	0.913	No
	Industry	4.74	62			
Teamwork	Faculty	4.68	40	0.064	0.258	No
	Industry	4.81	62			
Foreign language(s)	Faculty	4.71	41	0.618	0.810	No
	Industry	4.73	60			
Satisfaction						
Professionalism	Faculty	3.44	41	0.424	0.339	No
	Industry	3.24	55			
Ethics	Faculty	3.56	41	0.481	0.792	No
	Industry	3.62	55			
Adaptability	Faculty	3.24	41	0.837	0.954	No
	Industry	3.25	55			
Communication	Faculty	3.63	41	0.258	0.190	No
	Industry	3.34	56			
Teamwork	Faculty	3.76	41	0.734	0.593	No
	Industry	3.64	56			
Foreign language(s)	Faculty	3.63	41	0.691	0.853	No
	Industry	3.67	55			

Dimension IV: Business Management

See Table F.4.

Table F.4 The statistics of the students' evaluation of twenty-second-century skills importance currently and by 2030 their satisfaction level between faculty members and industrial sector perceptions "Dimension IV"

Dimension IV: Business management						
Skills	Mean		Sample number (N)	P value (Mann–Whitney U test)	P value (independent sample t test)	Statistical significance
Importance currently						
Management	Faculty	3.68	41	**0.037**	**0.054**	**Yes/No**
	Industry	4.05	62			
Leadership	Faculty	4.00	41	0.117	0.365	No

(continued)

Table F.4 (continued)

Dimension IV: Business management						
Skills	Mean		Sample number (N)	P value (Mann–Whitney U test)	P value (independent sample t test)	Statistical significance
			62			
Entrepreneurship	Faculty	3.85	41	0.844	0.995	No
	Industry	3.85	61			
Importance 2030						
Management	Faculty	4.24	41	0.079	0.059	No
	Industry	4.51	61			
Leadership	Faculty	4.41	41	0.095	0.090	No
	Industry	4.62	61			
Entrepreneurship	Faculty	4.37	41	0.778	0.696	No
	Industry	4.30	60			
Satisfaction						
Management	Faculty	2.98	40	0.903	0.910	No
	Industry	3.00	55			
Leadership	Faculty	3.18	40	0.981	0.975	No
	Industry	3.18	55			
Entrepreneurship	Faculty	2.85	40	0.510	0.507	No
	Industry	3.00	55			

Appendix G
Engineering Skills and Competencies Analysis Between Senior Students and Industry

See Tables G.1, G.2, G.3 and G.4.

Table G.1 The statistics of the students' evaluation of twenty-second-century skills importance currently and by 2030 their satisfaction level between senior students and industrial sector perceptions "Dimension I"

Dimension I: Core engineering knowledge and practice						
Skills	Mean		Sample number (N)	P value (Mann–Whitney U test)	P value (independent sample t test)	Statistical significance
Importance currently						
Disciplinary engineering fundamentals (depth)	Industry	4.44	61	**0.033**	**0.013**	**Yes**
	Senior students	4.09	287			
Interdisciplinary engineering fundamentals (breadth)	Industry	4.05	61	0.707	0.825	No
	Senior students	4.02	283			
Math, physics, and science fundamentals	Industry	4.15	59	0.758	0.433	No
	Senior students	4.03	281			
Practical experience	Industry	4.52	60	0.241	0.129	No
	Senior students	4.29	276			
ICT experience	Industry	4.36	59	0.201	0.068	No
	Senior students	4.07	286			
Multidisciplinary knowledge	Industry	3.74	61	**0.036**	0.115	**Yes/No**
	Senior students	3.98	281			

(continued)

M. Abdulwahed and M.O. Hasna, *Engineering and Technology Talent for Innovation and Knowledge-Based Economies*,
DOI 10.1007/978-3-319-46439-8

Table G.1 (continued)

Dimension I: Core engineering knowledge and practice						
Skills	Mean		Sample number (N)	P value (Mann–Whitney U test)	P value (independent sample t test)	Statistical significance
Importance 2030						
Disciplinary engineering fundamentals (depth)	Industry	4.62	60	**0.042**	**0.022**	**Yes**
	Senior students	4.30	288			
Interdisciplinary engineering fundamentals (breadth)	Industry	4.48	60	0.063	0.028	No/Yes
	Senior students	4.16	287			
Math, physics, and science fundamentals	Industry	4.28	58	0.311	0.134	No
	Senior students	4.03	288			
Practical experience	Industry	4.78	59	**0.044**	**0.013**	**Yes**
	Senior students	4.43	282			
ICT experience	Industry	4.66	58	**0.010**	**0.008**	**Yes**
	Senior students	4.24	288			
Multidisciplinary knowledge	Industry	4.10	60	0.552	0.962	No
	Senior students	4.09	281			
Disciplinary engineering fundamentals (depth)	Industry	3.65	55	0.323	0.414	No
	Senior students	3.53	287			
Interdisciplinary engineering fundamentals (breadth)	Industry	3.22	54	0.118	0.134	No
	Senior students	3.46	283			
Math, physics, and science fundamentals	Industry	3.60	53	0.395	0.738	No
	Senior students	3.66	288			
Practical experience	Industry	3.13	53	0.169	0.216	No
	Senior students	3.36	284			
ICT experience	Industry	3.76	54	0.078	0.064	No
	Senior students	3.47	287			
Multidisciplinary knowledge	Industry	3.13	53	0.129	0.110	No
	Senior students	3.40	286			

Table G.2 The statistics of the students' evaluation of twenty-second-century skills importance currently and by 2030 their satisfaction level between senior students and industrial sector perceptions "Dimension II"

Dimension II: Cognition and thinking						
Skills	Mean		Sample number (N)	P value (Mann–Whitney U test)	P value (independent sample t test)	Statistical significance
Importance currently						
Lifelong learning	Industry	4.46	61	0.189	0.110	No
	Senior students	4.23	278			
Problem-solving	Industry	4.43	61	0.539	0.575	No
	Senior students	4.35	283			
Decision-making	Industry	4.21	62	0.942	0.915	No
	Senior students	4.19	288			
Systems thinking	Industry	4.16	61	0.714	0.996	No
	Senior students	4.16	288			
Critical thinking	Industry	4.28	61	0.847	0.845	No
	Senior students	4.25	284			
Innovation	Industry	4.26	62	0.688	0.742	No
	Senior students	4.21	286			
Design	Industry	4.23	61	0.364	0.858	No
	Senior students	4.26	284			
Importance 2030						
Lifelong learning	Industry	4.72	60	**0.012**	**0.010**	**Yes**
	Senior students	4.34	280			
Problem-solving	Industry	4.62	60	0.582	0.147	No
	Senior students	4.41	287			
Decision-making	Industry	4.55	62	0.630	0.218	No
	Senior students	4.38	289			
Systems thinking	Industry	4.43	61	0.810	0.388	No
	Senior students	4.30	289			
Critical thinking	Industry	4.56	61	0.249	0.115	No
	Senior students	4.33	290			
Innovation	Industry	4.61	62	0.572	0.091	No
	Senior students	4.38	289			

(continued)

Table G.2 (continued)

Dimension II: Cognition and thinking						
Skills	Mean		Sample number (N)	P value (Mann–Whitney U test)	P value (independent sample t test)	Statistical significance
Design	Industry	4.57	60	0.837	0.203	No
	Senior students	4.37	288			
Satisfaction						
Lifelong learning	Industry	3.44	54	0.365	0.307	No
	Senior students	3.61	287			
Problem-solving	Industry	3.30	53	**0.005**	**0.013**	**Yes**
	Senior students	3.71	287			
Decision-making	Industry	3.22	55	**0.034**	**0.050**	**Yes**
	Senior students	3.53	288			
Systems thinking	Industry	3.42	55	**0.035**	**0.086**	**Yes/No**
	Senior students	3.68	286			
Critical thinking	Industry	3.18	55	**0.010**	**0.011**	**Yes**
	Senior students	3.58	287			
Innovation	Industry	3.11	55	0.069	0.050	No/Yes
	Senior students	3.43	285			
Design	Industry	3.35	55	**0.005**	**0.023**	**Yes**
	Senior students	3.72	284			

Table G.3 The statistics of the students' evaluation of twenty-second-century skills importance currently and by 2030 their satisfaction level between senior students and industrial sector perceptions "Dimension III"

Dimension III: Professional and interpersonal						
Skills	Mean		Sample number (N)	P value (Mann–Whitney U test)	P value (independent sample t test)	Statistical significance
Importance currently						
Professionalism	Industry	4.44	62	0.372	0.230	No
	Senior students	4.26	285			
Ethics	Industry	4.52	62	0.0482	0.268	No
	Senior students	4.35	284			

(continued)

Table G.3 (continued)

Dimension III: Professional and interpersonal						
Skills	Mean		Sample number (N)	P value (Mann–Whitney U test)	P value (independent sample t test)	Statistical significance
Adaptability	Industry	4.32	62	0.349	0.254	No
	Senior students	4.17	278			
Communication	Industry	4.53	62	0.176	0.132	No
	Senior students	4.31	281			
Teamwork	Industry	4.61	62	**0.047**	**0.044**	**Yes**
	Senior students	4.30	285			
Foreign language(s)	Industry	4.57	60	0.078	0.032	No/Yes
	Senior students	4.26	286			
Importance 2030						
Professionalism	Industry	4.69	61	0.082	0.032	No/Yes
	Senior students	4.40	285			
Ethics	Industry	4.76	62	0.078	0.021	No/Yes
	Senior students	4.43	288			
Adaptability	Industry	4.53	62	0.107	0.059	No
	Senior students	4.26	280			
Communication	Industry	4.74	62	**0.016**	**0.021**	**Yes**
	Senior students	4.40	284			
Teamwork	Industry	4.81	62	**0.005**	**0.006**	**Yes**
	Senior students	4.39	289			
Foreign language(s)	Industry	4.73	60	**0.006**	**0.003**	**Yes**
	Senior students	4.30	289			
Satisfaction						
Professionalism	Industry	3.24	55	**0.007**	**0.009**	**Yes**
	Senior students	3.66	286			
Ethics	Industry	3.62	55	0.452	0.572	No
	Senior students	3.72	284			
Adaptability	Industry	3.25	55	**0.004**	**0.008**	**Yes**
	Senior students	3.66	277			

(continued)

Table G.3 (continued)

Dimension III: Professional and interpersonal						
Skills	Mean		Sample number (N)	P value (Mann–Whitney U test)	P value (independent sample t test)	Statistical significance
Communication	Industry	3.34	56	**0.001**	**0.001**	**Yes**
	Senior students	3.87	286			
Teamwork	Industry	3.64	56	0.207	0.335	No
	Senior students	3.81	289			
Foreign language(s)	Industry	3.67	55	0.586	0.898	No
	Senior students	3.70	289			

Table G.4 The statistics of the students' evaluation of twenty-second-century skills importance currently and by 2030 their satisfaction level between senior students and industrial sector perceptions "Dimension IV"

Dimension IV: Business management						
Skills	Mean		Sample number (N)	P value (Mann–Whitney U test)	P value (independent sample t test)	Statistical significance
Importance currently						
Management	Industry	4.05	62	0.182	0.492	No
	Senior students	4.15	284			
Leadership	Industry	4.16	62	0.617	0.816	No
	Senior students	4.19	283			
Entrepreneurship	Industry	3.85	61	**0.010**	**0.049**	**Yes**
	Senior students	4.16	279			
Importance 2030						
Management	Industry	4.51	61	0.918	0.300	No
	Senior students	4.36	288			
Leadership	Industry	4.62	61	0.207	0.035	No/Yes
	Senior students	4.32	286			
Entrepreneurship	Industry	4.30	60	0.374	0.991	No
	Senior students	4.30	285			

(continued)

Table G.4 (continued)

Dimension IV: Business management						
Skills	Mean		Sample number (N)	P value (Mann–Whitney U test)	P value (independent sample t test)	Statistical significance
Satisfaction						
Management	Industry	3.00	55	**0.004**	**0.004**	**Yes**
	Senior students	3.46	280			
Leadership	Industry	3.18	55	**0.009**	**0.011**	**Yes**
	Senior students	3.60	281			
Entrepreneurship	Industry	3.00	55	**0.016**	**0.021**	**Yes**
	Senior students	3.40	277			

Appendix H
Engineering Skills and Competencies Analysis Between Males and Females

See Tables H.1, H.2, H.3 and H.4.

Table H.1 The statistics of the students' evaluation of twenty-second-century skills importance currently and by 2030 their satisfaction level between males and females sector perceptions "Dimension I"

Dimension I: Core engineering knowledge and practice						
Skills	Mean		Sample number (N)	P value (Mann–Whitney U test)	P value (independent sample t test)	Statistical significance
Importance currently						
Disciplinary engineering fundamentals (depth)	Male	107	3.96	0.077	0.113	No
	Female	180	4.17			
Interdisciplinary engineering fundamentals (breadth)	Male	105	3.85	0.051	0.035	No/Yes
	Female	178	4.12			
Math, physics, and science fundamentals	Male	105	3.98	0.465	0.551	No
	Female	176	4.06			
Practical experience	Male	104	4.27	0.863	0.772	No
	Female	172	4.31			
ICT experience	Male	106	3.98	0.363	0.291	No
	Female	180	4.13			
Multidisciplinary knowledge	Male	104	3.90	0.279	0.381	No
	Female	177	4.02			
Importance 2030						
Disciplinary engineering fundamentals (depth)	Male	106	4.26	0.476	0.700	No
	Female	182	4.31			

(continued)

M. Abdulwahed and M.O. Hasna, *Engineering and Technology Talent for Innovation and Knowledge-Based Economies*,
DOI 10.1007/978-3-319-46439-8

Table H.1 (continued)

Dimension I: Core engineering knowledge and practice						
Skills	Mean		Sample number (N)	P value (Mann–Whitney U test)	P value (independent sample t test)	Statistical significance
Interdisciplinary engineering fundamentals (breadth)	Male	107	3.96	0.016	0.021	Yes
	Female	180	4.27			
Math, physics, and science fundamentals	Male	106	3.97	0.340	0.513	No
	Female	182	4.07			
Practical experience	Male	104	4.40	0.485	0.793	No
	Female	178	4.44			
ICT experience	Male	105	4.10	0.047	0.093	Yes/No
	Female	183	4.33			
Multidisciplinary knowledge	Male	102	3.99	0.094	0.253	No
	Female	179	4.15			
Satisfaction						
Disciplinary engineering fundamentals (depth)	Male	105	3.49	0.664	0.549	No
	Female	182	3.56			
Interdisciplinary engineering fundamentals (breadth)	Male	103	3.42	0.550	0.645	No
	Female	180	3.48			
Math, physics, and science fundamentals	Male	103	3.75	0.499	0.284	No
	Female	185	3.61			
Practical experience	Male	104	3.26	0.310	0.300	No
	Female	180	3.42			
ICT experience	Male	104	3.41	0.483	0.500	No
	Female	183	3.50			
Multidisciplinary knowledge	Male	104	3.39	0.977	0.929	No
	Female	182	3.41			

Table H.2 The statistics of the students' evaluation of twenty-second-century skills importance currently and by 2030 their satisfaction level between males and females sector perceptions "Dimension II"

Dimension II: Cognition and thinking						
Skills	Mean		Sample number (N)	P value (Mann–Whitney U test)	P value (independent sample t test)	Statistical significance
Importance currently						
Lifelong learning	Male	103	4.13	0.316	0.207	No
	Female	175	4.29			
Problem-solving	Male	106	4.41	0.453	0.468	No
	Female	177	4.31			
Decision-making	Male	105	4.08	0.030	0.138	Yes/No
	Female	183	4.26			
Systems thinking	Male	105	3.99	0.017	0.035	Yes
	Female	183	4.26			
Critical thinking	Male	103	4.07	0.009	0.029	Yes
	Female	181	4.35			
Innovation	Male	105	4.00	0.007	0.012	Yes
	Female	181	4.33			
Design	Male	103	4.00	0.003	0.003	Yes
	Female	181	4.40			
Importance 2030						
Lifelong learning	Male	103	4.27	0.675	0.403	No
	Female	177	4.38			
Problem-solving	Male	106	4.39	0.728	0.765	No
	Female	181	4.43			
Decision-making	Male	106	4.24	0.011	0.066	Yes/No
	Female	183	4.46			
Systems thinking	Male	106	4.14	0.001	0.054	Yes/No
	Female	183	4.39			
Critical thinking	Male	107	4.23	0.105	0.260	No
	Female	183	4.38			
Innovation	Male	107	4.29	0.039	0.285	Yes/No
	Female	182	4.43			
Design	Male	106	4.23	0.014	0.114	Yes/No
	Female	182	4.45			
Satisfaction						
Lifelong learning	Male	104	3.69	0.294	0.324	No
	Female	183	3.56			
Problem-solving	Male	105	3.76	0.627	0.522	No
	Female	182	3.68			
Decision-making	Male	104	3.52	0.903	0.859	No
	Female	184	3.54			

(continued)

Table H.2 (continued)

Dimension II: Cognition and thinking						
Skills	Mean		Sample number (N)	P value (Mann–Whitney U test)	P value (independent sample t test)	Statistical significance
Systems thinking	Male	103	3.64	0.773	0.649	No
	Female	183	3.70			
Critical thinking	Male	104	3.52	0.384	0.450	No
	Female	183	3.62			
Innovation	Male	102	3.35	0.362	0.369	No
	Female	183	3.48			
Design	Male	101	3.51	0.026	0.023	Yes
	Female	183	3.84			

Table H.3 The statistics of the students' evaluation of twenty-second-century skills importance currently and by 2030 their satisfaction level between males and females sector perceptions "Dimension II"

Dimension III: Professional and interpersonal						
Skills	Mean		Sample number (N)	P value (Mann–Whitney U test)	P value (independent sample t test)	Statistical significance
Importance currently						
Professionalism	Male	105	4.11	0.072	0.083	No
	Female	180	4.34			
Ethics	Male	104	4.24	0.570	0.191	No
	Female	180	4.42			
Adaptability	Male	101	3.97	0.017	0.014	Yes
	Female	177	4.28			
Communication	Male	101	4.18	0.171	0.114	No
	Female	180	4.39			
Teamwork	Male	104	4.08	0.068	0.015	Yes
	Female	181	4.43			
Foreign language(s)	Male	104	4.08	0.004	0.024	Yes
	Female	182	4.37			
Importance 2030						
Professionalism	Male	105	4.35	0.215	0.578	No
	Female	180	4.42			
Ethics	Male	106	4.38	0.581	0.495	No
	Female	182	4.47			
Adaptability	Male	103	4.18	0.216	0.337	No
	Female	177	4.31			
Communication	Male	103	4.33	0.435	0.416	No
	Female	181	4.44			

(continued)

Table H.3 (continued)

Dimension III: Professional and interpersonal						
Skills	Mean		Sample number (N)	P value (Mann–Whitney U test)	P value (independent sample t test)	Statistical significance
Teamwork	Male	106	4.25	0.009	0.100	Yes/No
	Female	183	4.48			
Foreign language(s)	Male	106	4.25	0.267	0.517	No
	Female	183	4.33			
Satisfaction						
Professionalism	Male	103	3.41	0.003	0.003	Yes
	Female	183	3.80			
Ethics	Male	104	3.56	0.104	0.089	No
	Female	180	3.81			
Adaptability	Male	100	3.68	0.801	0.816	No
	Female	177	3.65			
Communication	Male	102	3.80	0.466	0.478	No
	Female	184	3.90			
Teamwork	Male	104	3.81	0.846	0.987	No
	Female	185	3.81			
Foreign language(s)	Male	104	3.56	0.213	0.154	No
	Female	185	3.77			

Table H.4 The statistics of the students' evaluation of twenty-second-century skills importance currently and by 2030 their satisfaction level between males and females sector perceptions "Dimension IV"

Dimension IV: Business management						
Skills	Mean		Sample number (N)	P value (Mann–Whitney U test)	P value (independent sample t test)	Statistical significance
Importance currently						
Management	Male	105	4.05	0.188	0.220	No
	Female	179	4.21			
Leadership	Male	103	4.00	0.008	0.015	Yes
	Female	180	4.31			
Entrepreneurship	Male	102	4.04	0.071	0.174	No
	Female	177	4.23			
Importance 2030						
Management	Male	105	4.27	0.194	0.227	No
	Female	183	4.42			
Leadership	Male	104	4.15	0.079	0.047	No/Yes
	Female	182	4.42			
Entrepreneurship	Male	102	4.25	0.154	0.548	No
	Female	183	4.33			

(continued)

Table H.4 (continued)

Dimension IV: Business management						
Skills	Mean		Sample number (N)	P value (Mann–Whitney U test)	P value (independent sample t test)	Statistical significance
Satisfaction						
Management	Male	100	3.34	0.310	0.154	No
	Female	180	3.53			
Leadership	Male	100	3.59	0.961	0.867	No
	Female	181	3.61			
Entrepreneurship	Male	97	3.55	0.110	0.141	No
	Female	180	3.33			

Appendix I
Engineering Skills and Competencies Analysis Between Nationals and Non-nationals

See Tables I.1, I.2, I.3 and I.4.

Table I.1 The statistics of the students' evaluation of twenty-second-century skills importance currently and by 2030 their satisfaction level between nationals and non-nationals sector perceptions "Dimension I"

Dimension I: Core engineering knowledge and practice						
Skills	Mean		Sample number (N)	P value (Mann–Whitney U test)	P value (independent sample t test)	Statistical significance
Importance currently						
Disciplinary engineering fundamentals (depth)	Qatari	85	4.05	0.707	0.592	No
	Non-Qatari	199	4.12			
Interdisciplinary engineering fundamentals (breadth)	Qatari	82	3.95	0.469	0.516	No
	Non-Qatari	198	4.04			
Math, physics, and science fundamentals	Qatari	83	3.88	0.189	0.126	No
	Non-Qatari	196	4.10			
Practical experience	Qatari	82	4.29	0.965	0.997	No
	Non-Qatari	191	4.29			
ICT experience	Qatari	85	4.06	0.938	0.882	No
	Non-Qatari	198	4.08			
Multidisciplinary knowledge	Qatari	83	4.04	0.704	0.593	No
	Non-Qatari	197	3.96			
Importance 2030						
Disciplinary engineering fundamentals (depth)	Qatari	86	4.22	0.724	0.411	No
	Non-Qatari	199	4.33			

(continued)

M. Abdulwahed and M.O. Hasna, *Engineering and Technology Talent for Innovation and Knowledge-Based Economies*,
DOI 10.1007/978-3-319-46439-8

Table I.1 (continued)

Dimension I: Core engineering knowledge and practice						
Skills	Mean		Sample number (N)	P value (Mann–Whitney U test)	P value (independent sample t test)	Statistical significance
Interdisciplinary engineering fundamentals (breadth)	Qatari	84	4.07	0.330	0.409	No
	Non-Qatari	200	4.19			
Math, physics, and science fundamentals	Qatari	87	3.91	0.319	0.229	No
	Non-Qatari	198	4.09			
Practical experience	Qatari	86	4.35	0.519	0.402	No
	Non-Qatari	194	4.46			
ICT experience	Qatari	86	4.19	0.875	0.543	No
	Non-Qatari	200	4.28			
Multidisciplinary knowledge	Qatari	85	4.18	0.257	0.437	No
	Non-Qatari	194	4.06			
Satisfaction						
Disciplinary engineering fundamentals (depth)	Qatari	86	3.53	0.839	0.966	No
	Non-Qatari	198	3.54			
Interdisciplinary engineering fundamentals (breadth)	Qatari	83	3.54	0.267	0.382	No
	Non-Qatari	197	3.42			
Math, physics, and science fundamentals	Qatari	86	3.53	0.292	0.213	No
	Non-Qatari	199	3.71			
Practical experience	Qatari	87	3.47	0.276	0.294	No
	Non-Qatari	194	3.30			
ICT experience	Qatari	87	3.43	0.838	0.631	No
	Non-Qatari	197	3.49			
Multidisciplinary knowledge	Qatari	86	3.62	0.030	0.03	Yes
	Non-Qatari	198	3.30			

Table I.2 The statistics of the students' evaluation of twenty-second-century skills importance currently and by 2030 their satisfaction level between nationals and non-nationals sector perceptions "Dimension II"

Dimension II: Cognition and thinking						
Skills	Mean	Sample number (N)		P value (Mann–Whitney U test)	P value (independent sample t test)	Statistical significance
Importance currently						
Lifelong learning	Qatari	83	4.23	0.804	0.969	No
	Non-Qatari	192	4.23			
Problem-solving	Qatari	84	4.29	0.473	0.470	No
	Non-Qatari	197	4.39			
Decision-making	Qatari	84	4.27	0.292	0.450	No
	Non-Qatari	202	4.17			
Systems thinking	Qatari	84	4.23	0.698	0.551	No
	Non-Qatari	201	4.14			
Critical thinking	Qatari	83	4.24	0.909	0.855	No
	Non-Qatari	199	4.27			
Innovation	Qatari	84	4.24	0.957	0.820	No
	Non-Qatari	199	4.21			
Design	Qatari	83	4.25	0.925	0.901	No
	Non-Qatari	199	4.27			
Importance 2030						
Lifelong learning	Qatari	84	4.32	0.977	0.809	No
	Non-Qatari	194	4.36			
Problem-solving	Qatari	87	4.34	0.485	0.454	No
	Non-Qatari	197	4.45			
Decision-making	Qatari	87	4.39	0.387	0.993	No
	Non-Qatari	199	4.39			
Systems thinking	Qatari	87	4.25	0.795	0.574	No
	Non-Qatari	200	4.33			

(continued)

Table I.2 (continued)

Dimension II: Cognition and thinking						
Skills	Mean	Sample number (N)		P value (Mann–Whitney U test)	P value (independent sample t test)	Statistical significance
Critical thinking	Qatari	87	4.26	0.811	0.491	No
	Non-Qatari	200	4.36			
Innovation	Qatari	87	4.31	0.568	0.456	No
	Non-Qatari	199	4.41			
Design	Qatari	86	4.27	0.697	0.317	No
	Non-Qatari	199	4.42			
Satisfaction						
Lifelong learning	Qatari	86	3.58	0.959	0.774	No
	Non-Qatari	198	3.62			
Problem-solving	Qatari	86	3.63	0.354	0.401	No
	Non-Qatari	198	3.75			
Decision-making	Qatari	88	3.48	0.621	0.595	No
	Non-Qatari	197	3.55			
Systems thinking	Qatari	88	3.65	0.817	0.769	No
	Non-Qatari	195	3.69			
Critical thinking	Qatari	88	3.63	0.531	0.665	No
	Non-Qatari	196	3.57			
Innovation	Qatari	88	3.51	0.374	0.421	No
	Non-Qatari	194	3.40			
Design	Qatari	87	3.75	0.861	0.789	No
	Non-Qatari	195	3.71			

Table I.3 The statistics of the students' evaluation of twenty-second-century skills importance currently and by 2030 their satisfaction level between nationals and non-nationals sector perceptions "Dimension III

Dimension III: Professional and interpersonal						
Skills	Mean		Sample number (N)	P value (Mann–Whitney U test)	P value (independent sample t test)	Statistical significance
Importance currently						
Professionalism	Qatari	83	4.29	0.885	0.817	No
	Non-Qatari	199	4.26			
Ethics	Qatari	83	4.27	0.230	0.318	No
	Non-Qatari	199	4.41			
Adaptability	Qatari	82	4.12	0.588	0.633	No
	Non-Qatari	194	4.19			
Communication	Qatari	82	4.20	0.091	0.185	No
	Non-Qatari	197	4.38			
Teamwork	Qatari	83	4.17	0.253	0.196	No
	Non-Qatari	200	4.37			
Foreign language(s)	Qatari	84	4.11	0.074	0.071	No
	Non-Qatari	200	4.35			
Importance 2030						
Professionalism	Qatari	85	4.27	0.556	0.165	No
	Non-Qatari	198	4.45			
Ethics	Qatari	86	4.30	0.294	0.171	No
	Non-Qatari	199	4.49			
Adaptability	Qatari	85	4.15	0.456	0.248	No
	Non-Qatari	192	4.31			
Communication	Qatari	85	4.26	0.426	0.156	No
	Non-Qatari	196	4.46			
Teamwork	Qatari	87	4.26	0.577	0.216	No
	Non-Qatari	199	4.45			
Foreign language(s)	Qatari	87	4.16	0.380	0.139	No
	Non-Qatari	199	4.37			
Satisfaction						
Professionalism	Qatari	86	3.69	0.764	0.796	No
	Non-Qatari	197	3.65			
Ethics	Qatari	86	3.73	0.935	0.874	No
	Non-Qatari	195	3.71			
Adaptability	Qatari	83	3.58	0.662	0.412	No
	Non-Qatari	191	3.69			
Communication	Qatari	87	3.79	0.373	0.446	No
	Non-Qatari	196	3.90			
Teamwork	Qatari	88	3.78	0.830	0.899	No
	Non-Qatari	198	3.80			
Foreign language(s)	Qatari	88	3.63	0.498	0.538	No
	Non-Qatari	198	3.72			

Table I.4 The statistics of the students' evaluation of twenty-second-century skills importance currently and by 2030 their satisfaction level between nationals and non-nationals sector perceptions "Dimension IV"

Dimension IV: Business management						
Skills	Mean		Sample number (N)	P value (Mann–Whitney U test)	P value (independent sample t test)	Statistical significance
Importance currently						
Management	Qatari	84	4.04	0.617	0.239	No
	Non-Qatari	197	4.20			
Leadership	Qatari	85	4.15	0.856	0.638	No
	Non-Qatari	195	4.22			
Entrepreneurship	Qatari	83	3.99	0.031	0.076	Yes/No
	Non-Qatari	193	4.24			
Importance 2030						
Management	Qatari	87	4.21	0.042	0.098	Yes/No
	Non-Qatari	198	4.43			
Leadership	Qatari	86	4.09	0.030	0.016	Yes
	Non-Qatari	197	4.43			
Entrepreneurship	Qatari	87	4.20	0.152	0.321	No
	Non-Qatari	195	4.34			
Satisfaction						
Management	Qatari	85	3.47	0.852	0.931	No
	Non-Qatari	192	3.46			
Leadership	Qatari	86	3.55	0.490	0.539	No
	Non-Qatari	192	3.64			
Entrepreneurship	Qatari	86	3.49	0.633	0.427	No
	Non-Qatari	188	3.37			

Appendix J
Twenty-First-Century Skills Situational Gaps for SEM Design Project

See Tables J.1, J.2, J.3 and J.4.

Table J.1 Perceptual gap of twenty-first-century skills 1st dimension between European, Asians, and Middle Eastern as well as genders comparing university studies and post-SEM participation

Dimension I: Core engineering knowledge and practice				
Variable	Group under study	Perceptual gap/Wilcoxon	Gender	Perceptual gap/Wilcoxon
Disciplinarily engineering fundamentals (depth)	European	0.69/0.000	Male	0.71/0.000
	Asian	0.66/0.000	Female	0.64/0.003
	Middle Eastern	0.72/0.001		
Interdisciplinary engineering knowledge (breadth)	European	0.76/0.000	Male	0.76/0.000
	Asian	0.75/0.000	Female	0.71/0.000
	Middle Eastern	0.71/0.001		
Math, physics, and science fundamentals	European	0.10/0.522	Male	0.18/0.121
	Asian	0.47/0.000	Female	0.06/0.972
	Middle Eastern	0.10/0.983		
Practical experience	European	1.42/0.000	Male	1.17/0.000
	Asian	0.87/0.000	Female	1.21/0.000
	Middle Eastern	1.12/0.000		
ICT experience	European	0.57/0.000	Male	0.45/0.000
	Asian	0.49/0.000	Female	0.77/0.000
	Middle Eastern	0.43/0.076		

(continued)

M. Abdulwahed and M.O. Hasna, *Engineering and Technology Talent for Innovation and Knowledge-Based Economies*,
DOI 10.1007/978-3-319-46439-8

Table J.1 (continued)

Dimension I: Core engineering knowledge and practice				
Variable	Group under study	Perceptual gap/Wilcoxon	Gender	Perceptual gap/Wilcoxon
Multidisciplinary knowledge	European	0.65/0.000	Male	0.68/0.000
	Asian	0.61/0.000	Female	0.57/0.001
	Middle Eastern	0.82/0.001		

Table J.2 Perceptual gap of twenty-first-century skills 2nd dimension between European, Asians, and Middle Eastern as well as genders comparing university studies and post-SEM participation

Dimension II: Cognition, thinking, and mental				
Variable	Group under study	Perceptual gap/Wilcoxon	Gender	Perceptual gap/Wilcoxon
Lifelong learning	European	0.4/0.000	Male	0.58/0.000
	Asian	0.66/0.000	Female	0.38/0.004
	Middle Eastern	0.44/0.131		
Problem-solving	European	0.76/0.000	Male	0.75/0.000
	Asian	0.58/0.000	Female	0.53/0.000
	Middle Eastern	0.85/0.001		
Decision-making	European	0.80/0.000	Male	0.80/0.000
	Asian	0.70/0.000	Female	0.67/0.001
	Middle Eastern	0.91/0.002		
Analytical thinking	European	0.30/0.000	Male	0.57/0.000
	Asian	0.78/0.000	Female	0.34/0.007
	Middle Eastern	0.58/0.080		
Systems thinking	European	0.49/0.000	Male	0.66/0.000
	Asian	0.78/0.000	Female	0.49/0.001
	Middle Eastern	0.69/0.005		
Critical thinking	European	0.60/0.000	Male	0.71/0.000
	Asian	0.71/0.000	Female	0.46/0.001
	Middle Eastern	0.65/0.022		
Creative thinking	European	0.87/0.000	Male	0.86/0.000
	Asian	0.79/0.000	Female	0.67/0.000
	Middle Eastern	0.71/0.005		

(continued)

Table J.2 (continued)

Dimension II: Cognition, thinking, and mental				
Variable	Group under study	Perceptual gap/Wilcoxon	Gender	Perceptual gap/Wilcoxon
Innovation	European	0.61/0.000	Male	0.72/0.000
	Asian	0.73/0.000	Female	0.53/0.001
	Middle Eastern	0.79/0.003		
Design	European	0.60/0.000	Male	0.71/0.000
	Asian	0.70/0.000	Female	0.50/0.002
	Middle Eastern	0.82/0.006		

Table J.3 Perceptual gap of twenty-first-century skills 3rd dimension between European, Asians, and Middle Eastern as well as genders comparing university studies and post-SEM participation

Dimension III: Professional and interpersonal				
Variable	Group under study	Perceptual gap/Wilcoxon	Gender	Perceptual gap/Wilcoxon
Professionalism	European	0.51/0.000	Male	0.57/0.000
	Asian	0.58/0.000	Female	0.59/0.000
	Middle Eastern	0.88/0.001		
Ethics and responsibility	European	0.38/0.000	Male	0.58/0.000
	Asian	0.72/0.000	Female	0.42/0.007
	Middle Eastern	0.64/0.010		
Adaptability	European	0.74/0.000	Male	0.73/0.000
	Asian	0.72/0.000	Female	0.64/0.001
	Middle Eastern	0.71/0.005		
Communication	European	0.68/0.000	Male	0.73/0.000
	Asian	0.69/0.000	Female	0.53/0.006
	Middle Eastern	0.71/0.004		
Teamwork	European	0.97/0.000	Male	0.95/0.000
	Asian	0.81/0.000	Female	0.77/0.000
	Middle Eastern	1.02/0.000		

Table J.4 Perceptual gap of twenty-first-century skills 4th dimension between European, Asians, and Middle Eastern as well as genders comparing university studies and post-SEM participation

Dimension IV: Business and management				
Variable	Group under study	Perceptual gap/Wilcoxon	Gender	Perceptual gap/Wilcoxon
Management	European	0.69/0.000	Male	0.73/0.000
	Asian	0.57/0.000	Female	0.41/0.007
	Middle Eastern	0.89/0.002		
Leadership	European	0.66/0.000	Male	0.76/0.000
	Asian	0.74/0.000	Female	0.48/0.010
	Middle Eastern	0.68/0.015		
Entrepreneurship	European	0.38/0.000	Male	0.6/0.000
	Asian	0.69/0.000	Female	0.35/0.009
	Middle Eastern	0.64/0.025		

Appendix K
Instruments

Senior Students Survey

Qatar University
College of Engineering
Students

This survey is conducted as part of large-scale study that focuses on preparedness of Engineering Graduates in Qatar currently and in light of Qatar National Vision (QNV) 2030. The survey is anonymous and your input is very essential in achieving the project objectives.

Section I – Background Information:

This section asks for some background information about you and your educational experience.

1. What is your gender, and nationality?
   O Male O Female | O Qatari O Non-Qatari, pls specify: ....

2. Please, specify in which university you are doing your undergraduate studies?
   O Qatar University O Texas A&M (Qatar) O College of North Atlantic Qatar
   O Carnegie Mellon university (Qatar) O Other: ....

3. Please, specify your engineering major? ....

4. What is your cumulative GPA: .... Total number of completed credit hours: ....

M. Abdulwahed and M.O. Hasna, *Engineering and Technology Talent for Innovation and Knowledge-Based Economies*,
DOI 10.1007/978-3-319-46439-8

Section II – Competencies & Skills of Engineering Graduates in Qatar:

This section asks for your views on the supply of suitably qualified engineering graduates direct from Universities in Qatar

5. The following are 21 sets of generic engineering graduate skills classified under 4 dimensions. Please indicate under "*Importance*" column **how much you perceive these skills to be important currently and in future for engineers (in light of QNV 2030)** (1= strongly disagree; 2= disagree; 3= Neutral; 4= agree; 5=strongly agree), and specify under the "*Satisfaction*" column **how much you are satisfied with the efficiency of undergraduate engineering education in preparing students with these skills to join the engineering practice** (1= strongly disagree; 2= disagree; 3= Neutral; 4= agree; 5=strongly agree)

Importance Currently 1 2 3 4 5	Importance 2030 1 2 3 4 5		Satisfaction 1 2 3 4 5
		Dimensions I: CORE ENGINEERING KNOWLEDGE & PRACTICE	
O O O O O	O O O O O	**DISCIPLINARILY ENGINEERING FUNDAMENTALS (Depth)** "Demonstrate in-depth technical knowledge in the engineering field of specialty and know-how in specific engineering discipline and apply this knowledge effectively through practice – in balance with non-technical skills"	O O O O O
O O O O O	O O O O O	**INTERDISCIPLINARY ENGINEERING KNOWLEDGE (Breadth)** "Develop and build a multidisciplinary knowledge from other engineering fields and a comprehensive interdisciplinary knowledge from a specific filed through a scholarly approach that incorporating theory and practice"	O O O O O
O O O O O	O O O O O	**MATH, PHYSICS, & SCIENCE FUNDAMENTALS** "Demonstrate essential mathematical skills and use, quantify, and manipulate numbers and their application accurately in any contexts"	O O O O O
O O O O O	O O O O O	**PRACTICAL EXPERIENCE** "Apply knowledge and skills, required for effective professional engineering practices, into real-world settings"	O O O O O
O O O O O	O O O O O	**ICT EXPERIENCE** "Use information and communication technology applications interactively in locating, managing, modeling, analyzing, evaluating, and processing information from multiple sources to effectively accomplish a given task"	O O O O O
O O O O O	O O O O O	**MULTIDISCIPLINARY KNOWLEDGE (Outside the engineering fields, e.g. social, humanities, business, etc.)** "Develop and build a multidisciplinary knowledge from other engineering fields and a comprehensive interdisciplinary knowledge from a specific filed through a scholarly approach that incorporating theory and practice"	O O O O O
		Dimensions II: COGNITION & THINKING	
O O O O O	O O O O O	**LIFELONG LEARNING** "Ability to continuously acquire new knowledge and skills for self and professional development at all levels (life and career)"	O O O O O
O O O O O	O O O O O	**PROBLEM SOLVING** "Use engineering knowledge systematically to identify, analyze, formulate, solve, and evaluate complex and multidisciplinary problems – by applying cognitive skills (logical, critical, and creative thinking)"	O O O O O

2

O O O O O	O O O O O	**DECISION MAKING** "Apply personal and professional judgment, take-risk and initiative, in effectively making strategic decision and managing risks, from a range of alternatives, based-on available information in response to ambiguous and complex situation"	O O O O O
O O O O O	O O O O O	**SYSTEMS THINKING** "Think systematically in holistic engineering approach by understanding how an entire system works and the relationship among systems' elements and other systems, and viewing a problem as a part of an overall system"	O O O O O
O O O O O	O O O O O	**CRITICAL THINKING** "Use critical, conceptual, reflective, and rational thinking in drawing and evidence-based assessing systematic conclusions and finding underlying relationships for solutions"	O O O O O
O O O O O	O O O O O	**INNOVATION** "Add values through introducing new novel ideas, methods, directions, opportunities, and solutions that meet new requirements, through more effective products, processes, services, and technologies, that are readily available to stakeholders"	O O O O O
O O O O O	O O O O O	**DESIGN** "Design engineering products to meet specified needs and develop optimal solutions for complex engineering problems"	O O O O O
		Dimensions III: PROFESSIONAL & INTERPERSONAL	
O O O O O	O O O O O	**PROFESSIONALISM** "Maintain high-level of social and professional behavior, demonstrate accountability, accept responsibility, and achieve excellence in work and everyday life"	O O O O O
O O O O O	O O O O O	**ETHICS** "Understand ethical norms and demonstrate ethical reasoning, behavior, and integrity of social responsibility and service in any context"	O O O O O
O O O O O	O O O O O	**ADAPTABILITY** "Demonstrate ingenuity in addressing challenges and flexibility and adaptability for actively managing rapid and diverse change successfully in workplace"	O O O O O
O O O O O	O O O O O	**COMMUNICATION** "Communicate effectively, in various ways (verbal, oral, written), with all stakeholders (Public, Engineers) across all boundaries (cultural, language)"	O O O O O
O O O O O	O O O O O	**TEAMWORK** "Work & cooperate efficiently in a diverse, multicultural, and interdisciplinary team"	O O O O O
O O O O O	O O O O O	**FOREIGN LANGUAGE(s)** "Demonstrate language proficiency, in foreign language(s) (in particular English for non-native speakers) at both level, professional and technical"	O O O O O

3

Dimensions IV: BUSINESS/ MANAGEMENT			
O O O O O	O O O O O	**MANAGEMENT** "Awareness of business literacy & practice in the context of engineering; manage organization & Resources efficiently, develop processes, plan projects, take-risk"	O O O O O
O O O O O	O O O O O	**LEADERSHIP** "Engaged, interact, lead, and influence people effectively toward accomplishing a common or shared goal and contributing positively to diverse communities"	O O O O O
O O O O O	O O O O O	**ENTREPRENEURSHIP** "Identify, capture, and develop an opportunity to create real business and social value through converting invention into successful innovation"	O O O O O
O O O O O	O O O O O	OTHER COMPETENCIES; please specify;	O O O O O

6. Please rank **top 5** engineering skills listed in question 5 above in order of importance (starting with 1 for the most important):

Rank 1 .... Rank 2 .... Rank 3 .... Rank 4 .... Rank 5 ....

7. Among the skills/competencies listed in question 5 above, which ones do you think are not adequately emphasized or should be more emphasized in undergraduate engineering education?

....

....

8. Do you think that competencies & skills demanded by Qatar industry would change from now till 2022 & from 2022 till 2030?

By 2022;	By 2030;
ONo O Yes	O No O Yes
Please provide further details for your response;	**Please provide further details for your response;**
....	....
....	....
....	....
....	....

4

Qatar University
College of Engineering
Students

9. **Which courses/ experiences have you been involved in during your undergraduate studies that has helped you the most in gaining and improving your skills**

Course/ experience	Skills improved/ gained	Please explain why?

Section III – Undergraduate Engineering Education in Qatar:

This section asks about engineering education system and its efficiency in preparing graduates for the work environment

10. **To what extent you agree that the following teaching styles/ learning approach are useful in preparing/qualifying students for the engineering practice? (1= strongly disagree; 2= disagree; 3= Neutral; 4= agree; 5=strongly agree)**

Teaching Style	1	2	3	4	5	Teaching Style	1	2	3	4	5
Lectures	O	O	O	O	O	Written exams	O	O	O	O	O
Group study	O	O	O	O	O	Instructor office hours	O	O	O	O	O
Research/ design projects	O	O	O	O	O	Tutorial sessions	O	O	O	O	O
Internship, training experience	O	O	O	O	O	Instructor feedback on your performance	O	O	O	O	O
Individual assignment	O	O	O	O	O	Class discussion	O	O	O	O	O
Group assignment	O	O	O	O	O	Online learning resources	O	O	O	O	O
Oral presentations	O	O	O	O	O	Labs	O	O	O	O	O
Practical Exams	O	O	O	O	O	Help and guidance from teaching assistances	O	O	O	O	O
Other..................	O	O	O	O	O	Other:	O	O	O	O	O

11. **To what extent do you agree or disagree that the following groups possess the following? (1= strongly disagree; 2= disagree; 3= Neutral; 4= agree; 5=strongly agree)**

	1	2	3	4	5	1	2	3	4	5	1	2	3	4	5
	Qatar university Students					Education City Students					Overseas Students				
Language proficiency	O	O	O	O	O	O	O	O	O	O	O	O	O	O	O
Competencies & skills	O	O	O	O	O	O	O	O	O	O	O	O	O	O	O
Knowledge (engineering & science)	O	O	O	O	O	O	O	O	O	O	O	O	O	O	O
Practical experience	O	O	O	O	O	O	O	O	O	O	O	O	O	O	O
Other:	O	O	O	O	O	O	O	O	O	O	O	O	O	O	O
Other:	O	O	O	O	O	O	O	O	O	O	O	O	O	O	O

Section IV – Training, practical experience, & Employment:

This section asks for information regarding your practical experience and organization of employment

12. **Have you been involved in internship, training experience during your undergraduate education?**

O No O Yes

If Yes, where and when did you take it? And did you find it beneficial in terms of competencies & skills?

Where? .... **When?**..

O beneficial O not beneficial

5

Qatar University
College of Engineering
Students

13. **What is your career plan after graduation?[check all that applies]**
    - ☐ Apply for postgraduate studies
    - ☐ Get employed
    - ☐ Have not decided yet
    - ☐ Start my own business
    - ☐ Stay at home
    - ☐ Other(s):.....................

14. **To what extent do you think that it is easy to get employed?**

O Strongly Disagree O Disagree O Neutral O Agree O Strongly Agree

If you disagree; to what extent do you agree with the following statements being a **potential barrier** from employment?
(1= Strongly Disagree, 2= Disagree, 3= Neutral, 4= Agree, 5=Strongly Agree)

Barriers	1	2	3	4	5	Barriers	1	2	3	4	5
The types of jobs available is not related to my engineering degree	O	O	O	O	O	I have family restrictions of getting employed	O	O	O	O	O
I am afraid of the new experience	O	O	O	O	O	My gender is a disadvantage	O	O	O	O	O
I am not interested in the types of jobs available in Qatar	O	O	O	O	O	I do not have work/ practical experience	O	O	O	O	O
I will continue my post graduate studies	O	O	O	O	O	My engineering knowledge is not adequate	O	O	O	O	O
My English is not professional enough	O	O	O	O	O	My nationality is a barrier	O	O	O	O	O
Graduates from my university are not hired easily	O	O	O	O	O	Others:....................	O	O	O	O	O

Section V – Comments & Further Suggestions:

15. **Please, add any further comments and/or recommendations you may consider to be important in helping local Engineering Colleges to prepare skilled engineers capable of meeting industry needs and governmental strategic plans in Qatar to fulfill the QNV 2030.**

....
....
....
....
....

Thanks for your participation in this survey...

Faculty Members Survey

Qatar University
College of Engineering
Faculty survey

This survey is conducted as part of large-scale study that focuses on preparedness of Engineering Graduates in Qatar currently and in the future in light of Qatar National Vision (QNV) 2030. The survey is anonymous and your input is very essential in achieving the project objectives.

Section I – Background Information:

This section asks for some background information and educational degrees.

1. What is your gender, and nationality?
○ Male ○ Female | ○ Qatari ○ Non-Qatari, pls specify: ……………

2. What is your educational background? [pls. check all that apply]

Degree	Major/ specialization	University	Year	GPA
☐ Bachelor (BSc)	……………………	…………	………	……
☐ Master (MSc) (if applicable)	……………………	…………	………	……
☐ Doctorate (PhD) (if applicable)	……………………	…………	………	……
☐ Other: …………	……………………	…………	………	……

3. Please, specify the undergraduate engineering program you are currently teaching?……………………

4. Please, specify the total number of years of your teaching experience? ……………………………

5. Currently, which of the following student group(s) are you teaching? [Pls. check all that apply]

☐ Freshman (0-29 credit hours) ☐ Sophomore (30-59 credit hours) ☐ Junior (60-89 credit hours)
☐ Senior (90+ credit hours) ☐ MSc ☐ PhD

6. Do you have a working experience in the industrial sector?
○ No ○ Yes
If Yes, for how many years? and please explain about the nature of your work?
…………………………………………………………………………………………………
…………………………………………………………………………………………………

7. Do you have professional experience outside Qatar? If Yes, which country(ies)
○ No ○ Yes, Country(ies)? ……………………………………

Section II – Competencies & Skills of Engineering Graduates in Qatar:

This section asks for your views on the supply of suitably qualified engineering graduates direct from Universities in Qatar

8. **The following are 21 sets of generic engineering graduate skills classified under 4 dimensions. Please indicate under** "*Importance*" **column how much you perceive these skills to be important currently and in future for engineers (in light of QNV 2030)** (1= strongly disagree; 2= disagree; 3= Neutral; 4= agree; 5=strongly agree), **and specify under the** "*Satisfaction*" **column how much you are satisfied with the efficiency of undergraduate engineering education in preparing students with these skills to join the engineering practice** (1= strongly disagree; 2= disagree; 3= Neutral; 4= agree; 5=strongly agree)

Importance Currently 1 2 3 4 5	Importance 2030 1 2 3 4 5		Satisfaction 1 2 3 4 5
		Dimensions I: CORE ENGINEERING KNOWLEDGE & PRACTICE	
O O O O O	O O O O O	**DISCIPLINARILY ENGINEERING FUNDAMENTALS (Depth)** "Demonstrate in-depth technical knowledge in the engineering field of specialty and know-how in specific engineering discipline and apply this knowledge effectively through practice – in balance with non-technical skills"	O O O O O
O O O O O	O O O O O	**INTERDISCIPLINARY ENGINEERING KNOWLEDGE (Breadth)** "Develop and build a multidisciplinary knowledge from other engineering fields and a comprehensive interdisciplinary knowledge from a specific filed through a scholarly approach that incorporating theory and practice"	O O O O O
O O O O O	O O O O O	**MATH, PHYSICS, & SCIENCE FUNDAMENTALS** "Demonstrate essential mathematical skills and use, quantify, and manipulate numbers and their application accurately in any contexts"	O O O O O
O O O O O	O O O O O	**PRACTICAL EXPERIENCE** "Apply knowledge and skills, required for effective professional engineering practices, into real-world settings"	O O O O O
O O O O O	O O O O O	**ICT EXPERIENCE** "Use information and communication technology applications interactively in locating, managing, modeling, analyzing, evaluating, and processing information from multiple sources to effectively accomplish a given task"	O O O O O
O O O O O	O O O O O	**MULTIDISCIPLINARY KNOWLEDGE (Outside the engineering fields, e.g. social, humanities, business, etc.)** "Develop and build a multidisciplinary knowledge from other engineering fields and a comprehensive interdisciplinary knowledge from a specific filed through a scholarly approach that incorporating theory and practice"	O O O O O
		Dimensions II: COGNITION & THINKING	
O O O O O	O O O O O	**LIFELONG LEARNING** "Ability to continuously acquire new knowledge and skills for self and professional development at all levels (life and career)"	O O O O O
O O O O O	O O O O O	**PROBLEM SOLVING** "Use engineering knowledge systematically to identify, analyze, formulate, solve, and evaluate complex and multidisciplinary problems – by applying cognitive skills (logical, critical, and creative thinking)"	O O O O O

2

O O O O O	O O O O O	**DECISION MAKING** "Apply personal and professional judgment, take-risk and initiative, in effectively making strategic decision and managing risks, from a range of alternatives, based-on available information in response to ambiguous and complex situation"	O O O O O
O O O O O	O O O O O	**SYSTEMS THINKING** "Think systematically in holistic engineering approach by understanding how an entire system works and the relationship among systems' elements and other systems, and viewing a problem as a part of an overall system"	O O O O O
O O O O O	O O O O O	**CRITICAL THINKING** "Use critical, conceptual, reflective, and rational thinking in drawing and evidence-based assessing systematic conclusions and finding underlying relationships for solutions"	O O O O O
O O O O O	O O O O O	**INNOVATION** "Add values through introducing new novel ideas, methods, directions, opportunities, and solutions that meet new requirements, through more effective products, processes, services, and technologies, that are readily available to stakeholders"	O O O O O
O O O O O	O O O O O	**DESIGN** "Design engineering products to meet specified needs and develop optimal solutions for complex engineering problems"	O O O O O
		Dimensions III: PROFESSIONAL & INTERPERSONAL	
O O O O O	O O O O O	**PROFESSIONALISM** "Maintain high-level of social and professional behavior, demonstrate accountability, accept responsibility, and achieve excellence in work and everyday life"	O O O O O
O O O O O	O O O O O	**ETHICS** "Understand ethical norms and demonstrate ethical reasoning, behavior, and integrity of social responsibility and service in any context"	O O O O O
O O O O O	O O O O O	**ADAPTABILITY** "Demonstrate ingenuity in addressing challenges and flexibility and adaptability for actively managing rapid and diverse change successfully in workplace"	O O O O O
O O O O O	O O O O O	**COMMUNICATION** "Communicate effectively, in various ways (verbal, oral, written), with all stakeholders (Public, Engineers) across all boundaries (cultural, language)"	O O O O O
O O O O O	O O O O O	**TEAMWORK** "Work & cooperate efficiently in a diverse, multicultural, and interdisciplinary team"	O O O O O
O O O O O	O O O O O	**FOREIGN LANGUAGE(s)** "Demonstrate language proficiency, in foreign language(s) (in particular English for non-native speakers) at both level, professional and technical"	O O O O O

Dimensions IV: BUSINESS/ MANAGEMENT			
O O O O O	O O O O O	**MANAGEMENT** "Awareness of business literacy & practice in the context of engineering, manage organization & Resources efficiently, develop processes, plan projects, take-risk"	O O O O O
O O O O O	O O O O O	**LEADERSHIP** "Engaged, interact, lead, and influence people effectively toward accomplishing a common or shared goal and contributing positively to diverse communities"	O O O O O
O O O O O	O O O O O	**ENTREPRENEURSHIP** "Identify, capture, and develop an opportunity to create real business and social value through converting invention into successful innovation"	O O O O O
O O O O O	O O O O O	OTHER COMPETENCIES, please specify, ...	O O O O O

9. **Please rank top 5 engineering skills listed in question 8 above in order of importance (starting with 1 for the most important):**

Rank 1 .... Rank 2 .... Rank 3 .... Rank 4 .... Rank 5 ....

10. **Among the skills/competencies listed in question 8 above, which ones do you think are not adequately emphasized or should be more emphasized in undergraduate engineering education?** ....

11. **To what extent do you agree or disagree that the following groups possess the following? (1= strongly disagree; 2= disagree; 3= Neutral; 4= agree; 5=strongly agree)**

	Male	Female	Qatari	Non- Qatari	Resident	Overseas	Qatar university	Other universities in Qatar
	1 2 3 4 5	1 2 3 4 5	1 2 3 4 5	1 2 3 4 5	1 2 3 4 5	1 2 3 4 5	1 2 3 4 5	1 2 3 4 5
Generic engineering skills	O O O O O	O O O O O	O O O O O	O O O O O	O O O O O	O O O O O	O O O O O	O O O O O
GPA	O O O O O	O O O O O	O O O O O	O O O O O	O O O O O	O O O O O	O O O O O	O O O O O
Work preparedness	O O O O O	O O O O O	O O O O O	O O O O O	O O O O O	O O O O O	O O O O O	O O O O O
Motivation	O O O O O	O O O O O	O O O O O	O O O O O	O O O O O	O O O O O	O O O O O	O O O O O
Management skills	O O O O O	O O O O O	O O O O O	O O O O O	O O O O O	O O O O O	O O O O O	O O O O O
Planning & organizing skills	O O O O O	O O O O O	O O O O O	O O O O O	O O O O O	O O O O O	O O O O O	O O O O O
Leadership skills	O O O O O	O O O O O	O O O O O	O O O O O	O O O O O	O O O O O	O O O O O	O O O O O

Qatar University
College of Engineering
Faculty survey

12. Do you think that competencies & skills demanded by Qatar industry would change from now till 2022 & from 2022 till 2030?

By 2022:	By 2030:
O No O Yes	O No O Yes
Please provide further details for your response;	**Please provide further details for your response;**
....	....
....	....
....	....
....	....
....	....

13. To what extent do you agree with the following statements being potential barriers from preparing students with the desired skills?(1=Strongly Disagree, 2=Disagree, 3=Neutral, 4=Agree, 5=Strongly Agree)

Barriers "student's side"	1	2	3	4	5	Barriers "collage side"	1	2	3	4	5
Language/ English proficiency	O	O	O	O	O	Faculty members don't have the required practical experience that gives the students a practical insight of their future profession	O	O	O	O	O
Students don't take their education seriously	O	O	O	O	O						
Deficiency in math & science skills	O	O	O	O	O	Facilities (like labs, equipment, classrooms, .. etc) are not enough since number of students is increasing	O	O	O	O	O
Lack of motivation	O	O	O	O	O	Collage doesn't focus on the professional side of engineering	O	O	O	O	O
Other:....................	O	O	O	O	O	Other:....................	O	O	O	O	O

Section III – Undergraduate Engineering Education in Qatar:

This section asks about engineering education system and its efficiency in preparing graduates for the work environment

14. To what extent do you agree/disagree that the following action by engineering education institutions will help better prepare graduates for the work place (1= Strongly Disagree, 2= Disagree, 3= Neutral, 4= Agree, 5= Strongly Agree)?

Statement	1	2	3	4	5
The university changes the curriculum depending on the needs of the country in the future	O	O	O	O	O
Increase collaboration with industry	O	O	O	O	O
Sector specific internship/work placements as an integral part of the engineering study program	O	O	O	O	O
Reward faculty members who conduct engineering education research	O	O	O	O	O
Faculty members think that there should be a professional practice course	O	O	O	O	O
Need of providing professional development office	O	O	O	O	O
Other:....................	O	O	O	O	O

5

Qatar University
College of Engineering
Faculty survey

Section IV – Industry- Academia Collaboration:

This section asks about your organization's practices regarding the linkages between Industry and Academia in Qatar.

15. **Does your organization have links with industry?**
   ○ No ○ Yes

If yes; for each of the following means of collaboration, please indicate through which means your university has collaborated and/or prefers to collaborate with industry. [Pls. check all that apply]

Means of Collaboration	Already collaborating				Would like to collaborate		
	Yes	No	If yes mention university name		Yes	No	If yes mention university name
Sponsorships for students	☐	☐	…………		☐	☐	…………
Engineering and career events	☐	☐	…………		☐	☐	…………
Graduate recruitment	☐	☐	…………		☐	☐	…………
Internships to graduates	☐	☐	…………		☐	☐	…………
Through 'real-life' projects and resources	☐	☐	…………		☐	☐	…………
University provides off-site workforce training for employee	☐	☐	…………		☐	☐	…………
Partner with industries for research and innovation	☐	☐	…………		☐	☐	…………
Others …………	☐	☐	…………		☐	☐	…………

16. **How much does your organization perceive collaboration with industry to be beneficial/ important for both sides?**
○ Very important ○ Important ○ Not important ○ Not important at all ○ N/A

17. **Do you think there are any barriers that limit interaction with industry?**
   ○ No ○ Yes

If yes; please indicate to which extent you agree with the following factors being a reason in limiting/preventing your organization from interacting with industry. [pls. check all that applies]

The university structure is not adapted to the needs of industrial collaborations	☐	There is no value or benefit to our university from such collaboration	☐
Inadequate laboratory facilities	☐	We had a bad experience with industry in the past	☐
Lack of autonomy to work with industry	☐	The process is too complicated	☐
Lack of organizational/ structural office/ unit to coordination	☐	No collaboration from industry side	☐
The collaboration is time consuming	☐	Difficulty fitting in with university timetable	☐
Inadequate infrastructure	☐	Other(s): …………	☐

Section V – Comments & Further Suggestions:

18. **Please, add any further comments and/or recommendations you may consider to be important in helping local Engineering Colleges to prepare skilled engineers capable of meeting industry needs and governmental strategic plans in Qatar to fulfill the QNV 2030.**

…………………………………………………………………………
…………………………………………………………………………
…………………………………………………………………………
…………………………………………………………………………
…………………………………………………………………………

Thanks for your participation in this survey…

6

Junior Engineers Survey

Qatar University
College of Engineering
Professional Engineers/ Alumni survey

This survey is conducted as part of large-scale study that focuses on preparedness of Engineering Graduates in Qatar currently and in light of Qatar National Vision (QNV) 2030. The survey is anonymous and your input is very essential in achieving the project objectives.

Section I – Background Information:

This section asks for some background information and educational degrees.

1. What is your gender, and nationality?
   - O Male O Female | O Qatari O Non-Qatari, pls specify: ……………

2. What is your educational background? [pls. check all that apply]

Degree	Major/ specialization	University	Year	GPA
☐ Diploma	……………………	…………	………	………
☐ Bachelor (BSc)	……………………	…………	………	………
☐ Master (MSc) (if applicable)	……………………	…………	………	………
☐ Doctorate (PhD) (if applicable)	……………………	…………	………	………
☐ Other: …………	……………………	…………	………	………

3. Please specify your employment status;
   - O I am currently employed and this is my first job since graduation
   - O I am currently employed and this is NOT my first job since graduation
   - O I am currently unemployed but I had at least one job since graduation
   - O I am currently unemployed and I did not have any job since graduation (skip question 4- 9)
   - O Others:……………

4. If you are currently employed, how many years of experience? ……………………

5. What is your primary job status?
   - O Employed as an engineer
   - O Employed as a technician
   - O Other: ……………………………
   - O Employed but NOT as an engineer NOR a technician, pls specify:…………
   - O Employed as a graduate/ research assistant

6. Please, specify your current organization of employment:……………………………………

7. Which of the following best describes your current employment role/function? And whether you are full-time or part-time employee?
   - O Research & Development
   - O Design
   - O Manufacturing/Production
   - O Sales and Procurement
   - O Project Management
   - O Cost analysis
   - O Troubleshooting
   - O Quality Control
   - O Health and safety
   - O Business/finance
   - O Consultant
   - O Human resources
   - O Management/Administration
   - O Marketing/Sales
   - O Other: ……………………

8. How much time would it take for an average graduate with the relevant educational background to become an expert in your current work?
   - O 1 to 6 Months O 6 to 12 Months O 12 to 18 Months O 18 to 24 Months O >24 Months

9. When did you get your first employment?
   - O Before graduation O 1- 6 months after graduation O 6-12 months after graduation O >12 months after graduation

1

Section II – Competencies & Skills of Engineering Graduates in Qatar:

This section asks for your views on the supply of suitably qualified engineering graduates direct from Universities in Qatar

10. **The following are 22 sets of generic engineering graduate skills classified under 4 dimensions. Please indicate under "*Importance*" column how much you perceive these skills to be important currently and in future for engineers (in light of QNV 2030)** (1= strongly disagree; 2= disagree; 3= Neutral; 4= agree; 5=strongly agree), **and specify under the "*Satisfaction*" column how much you are satisfied with the efficiency of undergraduate engineering education in preparing students with these skills to join the engineering practice** (1= strongly disagree; 2= disagree; 3= Neutral; 4= agree; 5=strongly agree)

Importance Currently 1 2 3 4 5	Importance 2030 1 2 3 4 5		Satisfaction 1 2 3 4 5
		Dimensions I: CORE ENGINEERING KNOWLEDGE & PRACTICE	
O O O O O	O O O O O	**DISCIPLINARILY ENGINEERING FUNDAMENTALS (Depth)** "Demonstrate in-depth technical knowledge in the engineering field of specialty and know-how in specific engineering discipline and apply this knowledge effectively through practice – in balance with non-technical skills"	O O O O O
O O O O O	O O O O O	**INTERDISCIPLINARY ENGINEERING KNOWLEDGE (Breadth)** "Develop and build a comprehensive interdisciplinary knowledge from a specific filed through a scholarly approach that incorporating theory and practice"	O O O O O
O O O O O	O O O O O	**MATH, PHYSICS, & SCIENCE FUNDAMENTALS** "Demonstrate essential mathematical skills and use, quantify, and manipulate numbers and their application accurately in any contexts"	O O O O O
O O O O O	O O O O O	**PRACTICAL EXPERIENCE** "Apply knowledge and skills, required for effective professional engineering practices, into real-world settings"	O O O O O
O O O O O	O O O O O	**ICT EXPERIENCE** "Use information and communication technology applications interactively in locating, managing, modeling, analyzing, evaluating, and processing information from multiple sources to effectively accomplish a given task"	O O O O O
O O O O O	O O O O O	**MULTIDISCIPLINARY KNOWLEDGE (Outside the engineering fields, e.g. social, humanities, business, etc.)** "Develop and build a multidisciplinary knowledge from other engineering fields through a scholarly approach that incorporating theory and practice"	O O O O O
		Dimensions II: COGNITION & THINKING	
O O O O O	O O O O O	**LIFELONG LEARNING** "Ability to continuously acquire new knowledge and skills for self and professional development at all levels (life and career)"	O O O O O
O O O O O	O O O O O	**PROBLEM SOLVING** "Use engineering knowledge systematically to identify, analyze, formulate, solve, and evaluate complex and multidisciplinary problems – by applying cognitive skills (logical, critical, and creative thinking)"	O O O O O

2

Importance Currently 1 2 3 4 5	Importance 2030 1 2 3 4 5		Satisfaction 1 2 3 4 5
O O O O O	O O O O O	**DECISION MAKING** "Apply personal and professional judgment, take-risk and initiative, in effectively making strategic decision and managing risks, from a range of alternatives, based-on available information in response to ambiguous and complex situation"	O O O O O
O O O O O	O O O O O	**SYSTEMS THINKING** "Think systematically in holistic engineering approach by understanding how an entire system works and the relationship among systems' elements and other systems, and viewing a problem as a part of an overall system"	O O O O O
O O O O O	O O O O O	**CRITICAL THINKING** "Use critical, conceptual, reflective, and rational thinking in drawing and evidence-based assessing systematic conclusions and finding underlying relationships for solutions"	O O O O O
O O O O O	O O O O O	**INNOVATION** "Add values through introducing new novel ideas, methods, directions, opportunities, and solutions that meet new requirements, through more effective products, processes, services, and technologies, that are readily available to stakeholders"	O O O O O
O O O O O	O O O O O	**DESIGN** "Design engineering products to meet specified needs and develop optimal solutions for complex engineering problems"	O O O O O
		Dimensions III: PROFESSIONAL & INTERPERSONAL	
O O O O O	O O O O O	**PROFESSIONALISM** "Maintain high-level of social and professional behavior, demonstrate accountability, accept responsibility, and achieve excellence in work and everyday life"	O O O O O
O O O O O	O O O O O	**ETHICS** "Understand ethical norms and demonstrate ethical reasoning, behavior, and integrity of social responsibility and service in any context"	O O O O O
O O O O O	O O O O O	**ADAPTABILITY** "Demonstrate ingenuity in addressing challenges and flexibility and adaptability for actively managing rapid and diverse change successfully in workplace"	O O O O O
O O O O O	O O O O O	**COMMUNICATION** "Communicate effectively, in various ways (verbal, oral, written), with all stakeholders (Public, Engineers) across all boundaries (cultural, language)"	O O O O O
O O O O O	O O O O O	**TEAMWORK** "Work & cooperate efficiently in a diverse, multicultural, and interdisciplinary team"	O O O O O
O O O O O	O O O O O	**FOREIGN LANGUAGE(s)** "Demonstrate language proficiency, in foreign language(s) (in particular English for non-native speakers) at both level, professional and technical"	O O O O O

3

Importance Currently 1 2 3 4 5	Importance 2030 1 2 3 4 5		Satisfaction 1 2 3 4 5
		Dimensions IV: BUSINESS/ MANAGEMENT	
O O O O O	O O O O O	**MANAGEMENT** "Awareness of business literacy & practice in the context of engineering, manage organization & Resources efficiently, develop processes, plan projects, take-risk"	O O O O O
O O O O O	O O O O O	**LEADERSHIP** "Engaged, interact, lead, and influence people effectively toward accomplishing a common or shared goal and contributing positively to diverse communities"	O O O O O
O O O O O	O O O O O	**ENTREPRENEURSHIP** "Identify, capture, and develop an opportunity to create real business and social value through converting invention into successful innovation"	O O O O O
O O O O O	O O O O O	OTHER COMPETENCIES, please specify;	O O O O O

11. **Please rank top 5 engineering skills listed in question 10 above in order of importance (starting with 1 for the most important):**

Rank 1 .... Rank 2 .... Rank 3 .... Rank 4 .... Rank 5 ....

12. **Among the skills/competencies listed in question 10 above, which ones do you think are not adequately emphasized or should be more emphasized in undergraduate engineering education?**

....

....

13. **Do you think that competencies & skills demanded by Qatar industry would change from now till 2022 & from 2022 till 2030?**

By 2022:	By 2030:
O No O Yes	O No O Yes
Please provide further details for your response;	**Please provide further details for your response;**
....	....
....	....
....	....
....	....

4

Qatar University
College of Engineering
Professional Engineers/ Alumni survey

14. **Which courses/ experiences have you been involved in during your undergraduate studies that has helped you the most in gaining and improving your skills**

Course/ experience	**Skills improved/ gained**	**Please explain why?**

Section III – Undergraduate Engineering Education in Qatar:

This section asks about engineering education system and its efficiency in preparing graduates for the work environment

15. **Looking back, if you were free to choose again would you choose the same undergraduate study program(s) at the same collage?**

O Yes
O No, the same study program at a different collage
O No, I would decide not to study at all
O No, a different study program at the same collage
O No, a different study program at a different collage
O Other: ....

16. **To what extent you agree that the following teaching styles/ learning approach are useful in preparing/qualifying students for the engineering practice?** (1= strongly disagree; 2= disagree; 3= Neutral; 4= agree; 5=strongly agree)

Teaching Style	1	2	3	4	5	Teaching Style	1	2	3	4	5
Lectures	O	O	O	O	O	Written exams	O	O	O	O	O
Group study	O	O	O	O	O	Instructor office hours	O	O	O	O	O
Research / design projects	O	O	O	O	O	Tutorial sessions	O	O	O	O	O
Internship, training experience	O	O	O	O	O	Instructor feedback on your performance	O	O	O	O	O
Individual assignment	O	O	O	O	O	Class discussion	O	O	O	O	O
Group assignment	O	O	O	O	O	Online learning resources	O	O	O	O	O
Oral presentations	O	O	O	O	O	Labs	O	O	O	O	O
Practical Exams	O	O	O	O	O	Help and guidance from teaching assistances	O	O	O	O	O
Other:..............................	O	O	O	O	O	Other:	O	O	O	O	O

17. **To what extent do you agree or disagree that the following groups possess the following?** (1= strongly disagree; 2= disagree; 3= Neutral; 4= agree; 5=strongly agree)

	1	2	3	4	5	1	2	3	4	5	1	2	3	4	5
	Qatar university Students					**Education City Students**					**Overseas Students**				
Language proficiency	O	O	O	O	O	O	O	O	O	O	O	O	O	O	O
Competencies & skills	O	O	O	O	O	O	O	O	O	O	O	O	O	O	O
Knowledge (engineering & science)	O	O	O	O	O	O	O	O	O	O	O	O	O	O	O
Practical experience	O	O	O	O	O	O	O	O	O	O	O	O	O	O	O
Other:	O	O	O	O	O	O	O	O	O	O	O	O	O	O	O
Other:	O	O	O	O	O	O	O	O	O	O	O	O	O	O	O

Qatar University
College of Engineering
Professional Engineers/ Alumni survey

Section IV – Training, practical experience, & Employment:

This section asks for information regarding your practical experience and organization of employment.

18. **Have you been involved in internship, training experience during your undergraduate education?**
O No O Yes

If Yes, where and when did you take it? And did you find it beneficial in terms of competencies & skills?

Where? .... **When?**...

O **beneficial** O **not beneficial**

19. **To what extent do you think that it is easy to get employed?**

O Strongly Disagree O Disagree O Neutral O Agree O Strongly Agree

If you disagree; to what extent do you agree with the following statements being a potential barrier from employment?
(1= Strongly Disagree, 2= Disagree, 3= Neutral, 4= Agree, 5=Strongly Agree)

Barriers	1	2	3	4	5	Barriers	1	2	3	4	5
The types of jobs available is not related to my engineering degree	O	O	O	O	O	I have family restrictions of getting employed	O	O	O	O	O
I am afraid of the new experience	O	O	O	O	O	My gender is a disadvantage	O	O	O	O	O
I am not interested in the types of jobs available in Qatar	O	O	O	O	O	I do not have work/ practical experience	O	O	O	O	O
I will continue my post graduate studies	O	O	O	O	O	My engineering knowledge is not adequate	O	O	O	O	O
My English is not professional enough	O	O	O	O	O	My nationality is a barrier	O	O	O	O	O
Graduates from my university are not hired easily	O	O	O	O	O	Others..........................	O	O	O	O	O

Section V – Comments & Further Suggestions:

20. **Please, add any further comments and/or recommendations you may consider to be important in helping local Engineering Colleges to prepare skilled engineers capable of meeting industry needs and governmental strategic plans in Qatar to fulfill the QNV 2030.**

....
....
....
....
....

Thanks for your participation in this survey...

Managers Surveys

Qatar University
College of Engineering
Managers survey

This survey is conducted as part of large-scale study that focuses on preparedness of Engineering Graduates in Qatar currently and in light of Qatar National Vision (QNV) 2030. The survey is anonymous and your input is very essential in achieving the project objectives.

Section I – Background Information:

This section asks for some background information about you and your organization.

1. **What is your gender, and nationality?**
O Male O Female | O Qatari O Non-Qatari, pls specify............... | Organization:..................

2. **What is your highest educational degree, which specialization, which university, and when did you obtain it?**..
....

3. **Which of the following best describes your position at work? [Pls. check all that apply].**

☐CEO/ COO/ President ☐ Owner/ Founder ☐Vice-president ☐ Director
☐Manager/Supervisor ☐ Senior Engineer ☐Training Specialist ☐ Other(s): ....

Section II – Competencies & Skills of Engineering Graduates in Qatar:

This section asks for your views on the supply of suitably qualified engineering graduates direct from universities in Qatar

4. **The following are 22 sets of generic engineering graduate skills classified under 4 dimensions. Please indicate under** "*Importance*" **column how much you perceive these skills to be important currently and in future for engineers (in light of QNV 2030)** (1= strongly disagree; 2= disagree; 3= Neutral; 4= agree; 5=strongly agree), **and specify under the** "*Satisfaction*" **column how much you are satisfied with the efficiency of undergraduate engineering education in preparing students with these skills to join the engineering practice** (1= strongly disagree; 2= disagree; 3= Neutral; 4= agree; 5=strongly agree)

Importance Currently 1 2 3 4 5	Importance 2030 1 2 3 4 5		Satisfaction 1 2 3 4 5
		Dimensions I: CORE ENGINEERING KNOWLEDGE & PRACTICE	
O O O O O	O O O O O	**DISCIPLINARILY ENGINEERING FUNDAMENTALS (Depth)** "Demonstrate in-depth technical knowledge in the engineering field of specialty and know-how in specific engineering discipline and apply this knowledge effectively through practice – in balance with non-technical skills"	O O O O O
O O O O O	O O O O O	**INTERDISCIPLINARY ENGINEERING KNOWLEDGE (Breadth)** "Develop and build a comprehensive interdisciplinary knowledge from a specific filed through a scholarly approach that incorporating theory and practice"	O O O O O
O O O O O	O O O O O	**MATH, PHYSICS, & SCIENCE FUNDAMENTALS** "Demonstrate essential mathematical skills and use, quantify, and manipulate numbers and their application accurately in any contexts"	O O O O O
O O O O O	O O O O O	**PRACTICAL EXPERIENCE** "Apply knowledge and skills, required for effective professional engineering practices, into real-world settings"	O O O O O
O O O O O	O O O O O	**ICT EXPERIENCE** "Use information and communication technology applications interactively in locating, managing, modeling, analyzing, evaluating, and processing information from multiple sources to effectively accomplish a given task"	O O O O O
O O O O O	O O O O O	**MULTIDISCIPLINARY KNOWLEDGE (Outside the engineering fields, e.g. social, humanities, business, etc.)** "Develop and build a multidisciplinary knowledge from other engineering fields through a scholarly approach that incorporating theory and practice"	O O O O O
		Dimensions II: COGNITION & THINKING	
O O O O O	O O O O O	**LIFELONG LEARNING** "Ability to continuously acquire new knowledge and skills for self and professional development at all levels (life and career)"	O O O O O
O O O O O	O O O O O	**PROBLEM SOLVING** "Use engineering knowledge systematically to identify, analyze, formulate, solve, and evaluate complex and multidisciplinary problems – by applying cognitive skills (logical, critical, and creative thinking)"	O O O O O

2

Importance Currently 1 2 3 4 5	Importance 2030 1 2 3 4 5		Satisfaction 1 2 3 4 5
O O O O O	O O O O O	**DECISION MAKING** "Apply personal and professional judgment, take-risk and initiative, in effectively making strategic decision and managing risks, from a range of alternatives, based-on available information in response to ambiguous and complex situation"	O O O O O
O O O O O	O O O O O	**SYSTEMS THINKING** "Think systematically in holistic engineering approach by understanding how an entire system works and the relationship among systems' elements and other systems, and viewing a problem as a part of an overall system"	O O O O O
O O O O O	O O O O O	**CRITICAL THINKING** "Use critical, conceptual, reflective, and rational thinking in drawing and evidence-based assessing systematic conclusions and finding underlying relationships for solutions"	O O O O O
O O O O O	O O O O O	**INNOVATION** "Add values through introducing new novel ideas, methods, directions, opportunities, and solutions that meet new requirements, through more effective products, processes, services, and technologies, that are readily available to stakeholders"	O O O O O
O O O O O	O O O O O	**DESIGN** "Design engineering products to meet specified needs and develop optimal solutions for complex engineering problems"	O O O O O
		Dimensions III: PROFESSIONAL & INTERPERSONAL	
O O O O O	O O O O O	**PROFESSIONALISM** "Maintain high-level of social and professional behavior, demonstrate accountability, accept responsibility, and achieve excellence in work and everyday life"	O O O O O
O O O O O	O O O O O	**ETHICS** "Understand ethical norms and demonstrate ethical reasoning, behavior, and integrity of social responsibility and service in any context"	O O O O O
O O O O O	O O O O O	**ADAPTABILITY** "Demonstrate ingenuity in addressing challenges and flexibility and adaptability for actively managing rapid and diverse change successfully in workplace"	O O O O O
O O O O O	O O O O O	**COMMUNICATION** "Communicate effectively, in various ways (verbal, oral, written), with all stakeholders (Public, Engineers) across all boundaries (cultural, language)"	O O O O O
O O O O O	O O O O O	**TEAMWORK** "Work & cooperate efficiently in a diverse, multicultural, and interdisciplinary team"	O O O O O
O O O O O	O O O O O	**FOREIGN LANGUAGE(s)** "Demonstrate language proficiency, in foreign language(s) (in particular English for non-native speakers) at both level, professional and technical"	O O O O O

3

Importance Currently 1 2 3 4 5	Importance 2030 1 2 3 4 5		Satisfaction 1 2 3 4 5
		Dimensions IV: BUSINESS/ MANAGEMENT	
O O O O O	O O O O O	**MANAGEMENT** "Awareness of business literacy & practice in the context of engineering, manage organization & Resources efficiently, develop processes, plan projects, take-risk"	O O O O O
O O O O O	O O O O O	**LEADERSHIP** "Engaged, interact, lead, and influence people effectively toward accomplishing a common or shared goal and contributing positively to diverse communities"	O O O O O
O O O O O	O O O O O	**ENTREPRENEURSHIP** "Identify, capture, and develop an opportunity to create real business and social value through converting invention into successful innovation"	O O O O O
O O O O O	O O O O O	OTHER COMPETENCIES, please specify,............	O O O O O

5. **Please rank top 5 engineering skills listed in question 4 above in order of importance (starting with 1 for the most important):**

Rank 1 .... Rank 2 .... Rank 3 .... Rank 4 .... Rank 5 ....

6. **Among the skills/competencies listed in question 4 above, which ones do you think are not adequately emphasized or should be more emphasized in undergraduate engineering education?**..

....

7. **To what extent do you agree or disagree that the following groups possess the following? (1= strongly disagree; 2= disagree; 3= Neutral; 4= agree; 5=strongly agree)**

	Male 1 2 3 4 5	Female 1 2 3 4 5	Qatari 1 2 3 4 5	Non- Qatari 1 2 3 4 5	Resident 1 2 3 4 5	Overseas 1 2 3 4 5	Qatar university 1 2 3 4 5	Other universities in Qatar 1 2 3 4 5
Generic engineering skills	O O O O O	O O O O O	O O O O O	O O O O O	O O O O O	O O O O O	O O O O O	O O O O O
GPA	O O O O O	O O O O O	O O O O O	O O O O O	O O O O O	O O O O O	O O O O O	O O O O O
Work preparedness	O O O O O	O O O O O	O O O O O	O O O O O	O O O O O	O O O O O	O O O O O	O O O O O
Motivation	O O O O O	O O O O O	O O O O O	O O O O O	O O O O O	O O O O O	O O O O O	O O O O O
Management skills	O O O O O	O O O O O	O O O O O	O O O O O	O O O O O	O O O O O	O O O O O	O O O O O
Planning & organizing skills	O O O O O	O O O O O	O O O O O	O O O O O	O O O O O	O O O O O	O O O O O	O O O O O
Leadership skills	O O O O O	O O O O O	O O O O O	O O O O O	O O O O O	O O O O O	O O O O O	O O O O O

4

Qatar University
College of Engineering
Managers survey

8. To which extent do you agree that it is **important** for the following groups to master the following **Management and Business competencies** for their professional development in your organization? (1= strongly disagree; 2= disagree; 3= Neutral; 4= agree; 5=strongly agree)

	Male 1	2	3	4	5	Female 1	2	3	4	5	Qatari 1	2	3	4	5	Non-Qatari 1	2	3	4	5
Project management skills	O	O	O	O	O	O	O	O	O	O	O	O	O	O	O	O	O	O	O	O
Risk management skills	O	O	O	O	O	O	O	O	O	O	O	O	O	O	O	O	O	O	O	O
Planning & organizing skills	O	O	O	O	O	O	O	O	O	O	O	O	O	O	O	O	O	O	O	O
Leadership skills	O	O	O	O	O	O	O	O	O	O	O	O	O	O	O	O	O	O	O	O

9. Do you think that competencies & skills demanded by Qatar industry would change from now till 2022 & from 2022 till 2030?

By 2022:	By 2030:
O No O Yes	O No O Yes
Please provide further details for your response;	**Please provide further details for your response;**
....	....
....	....
....	....
....	....
....	....

Section III – Recruitment & Employment

This section asks for information regarding recruitment and employment in your organization.

10. Please indicate your organizations degree of interest in recruiting engineers; (1= no interest; 2= low interest; 3= Neutral ; 4= moderate interest; 5= considerable interest)

	1	2	3	4	5		1	2	3	4	5		1	2	3	4	5
Fresh graduate	O	O	O	O	O	**Qatari**	O	O	O	O	O	**Qatar university**	O	O	O	O	O
Experienced (5+ years)	O	O	O	O	O	**Non-Qatari**	O	O	O	O	O	**Other universities**	O	O	O	O	O
												Pls specify:.........					
Male	O	O	O	O	O	**Resident**	O	O	O	O	O	**BSc**	O	O	O	O	O
Female	O	O	O	O	O	**Overseas**	O	O	O	O	O	**MSc**	O	O	O	O	O
												PhD	O	O	O	O	O

11. How much is the average projected increase/decrease in employing new engineering staff PER-YEAR over the next three to five years?

O Increase (Per-Year%): ....%
O Decrease (Per-Year%): ....%
O No Increase or Decrease (Same as current)
O NA (No Idea or Projection is available)

Qatar University
College of Engineering
Managers survey

12. Does your organization recruit engineering graduates who are **not** Qatar residents?
O No O Yes
If yes; what is the reason? [Pls. check all that apply]

☐They are hardworking and has better performance
☐They are willing to work for less
☐Recruiting the very best talent available
☐Their disciplinary engineering field is not found in Qatar
☐Other: ……………………………
☐Insufficient candidates with the right skills in Qatar
☐More experienced than graduates from local universities
☐They have the needed skills
☐They have better work professionalism

13. What are the engineering disciplines that fill in your organization **now**? & what are the disciplines you expect to recruit to satisfy the needs and the plans of your company over the **next 5-15 years** (till 2022 & 2030)? And what is the **estimated percentage** of each discipline? [Pls. check all that apply]

Engineering Discipline	Now	Estimated % now	Till 2022	Estimated % till 2022	2022-2030	Estimated % 2022-2030
Architecture Engineering	☐	…%	☐	…%	☐	…%
Biomedical Engineering	☐	…%	☐	…%	☐	…%
Civil Engineering	☐	…%	☐	…%	☐	…%
Chemical Engineering	☐	…%	☐	…%	☐	…%
Computer/ software Engineering	☐	…%	☐	…%	☐	…%
Electrical Engineering	☐	…%	☐	…%	☐	…%
Automation/Control	☐	…%	☐	…%	☐	…%
Industrial & systems Engineering	☐	…%	☐	…%	☐	…%
Mechanical Engineering	☐	…%	☐	…%	☐	…%
Petroleum Engineering	☐	…%	☐	…%	☐	…%
Other(s): ………………	☐	…%	☐	…%	☐	…%

Section IV – Industry- Academia Collaboration:

This section asks about your organization's practices regarding the linkages between Industry and Academia in Qatar.

14. Does your organization have links with engineering universities/colleges?
O No O Yes
If yes; for each of the following means of collaboration, please indicate through which means your company has collaborated and/or prefers to collaborate with universities. [Pls. check all that apply]

Means of Collaboration	Already collaborating				Would like to collaborate		
	Yes	No	If yes mention university name		Yes	No	If yes mention university name
Sponsorships for students	☐	☐	………………………		☐	☐	………………………
Engineering and career events	☐	☐	………………………		☐	☐	………………………
Graduate recruitment	☐	☐	………………………		☐	☐	………………………
Offer internships to graduates	☐	☐	………………………		☐	☐	………………………
Provide 'real-life' projects and resources	☐	☐	………………………		☐	☐	………………………
University provides off-site workforce training for our employee	☐	☐	………………………		☐	☐	………………………
Partner with universities for research and innovation	☐	☐	………………………		☐	☐	………………………
Others ………………	☐	☐	………………………		☐	☐	………………………

15. How much does your organization perceive collaboration with universities to be beneficial/ important for both sides?
O Very important O Important O Not important O Not important at all O N/A

Qatar University
College of Engineering
Managers survey

16. **Do you think there are any barriers that limit interaction with universities?**
O No O Yes

If yes; please indicate to which extent you agree with the following factors being a reason in limiting/preventing your organization from interacting with Universities. [pls. check all that applies]

Inadequate infrastructure	☐	There is no value or benefit to our company from such collaboration	☐
Inadequate laboratory facilities	☐	We had a bad experience with university in the past	☐
Lack of autonomy to work with academia	☐	The process is too complicated	☐
The university structure is not adapted to the needs of industrial collaborations	☐	Local universities are not interested/ no collaboration from university side	☐
The collaboration is time consuming	☐	Difficulty fitting in with university timetable	☐
Lack of organizational/ structural office/ unit to coordination	☐	Other(s):	☐

Section V – Comments & Further Suggestions:

17. **Please, add any further comments and/or recommendations you may consider to be important in helping local Engineering Colleges to prepare skilled engineers capable of meeting industry needs and governmental strategic plans in Qatar to fulfill the QNV 2030.**

....
....
....
....
....

Thanks for your participation in this survey...

HR Survey

Qatar University
College of Engineering
HR survey

This survey is conducted as part of large-scale study that focuses on preparedness of Engineering Graduates in Qatar currently and in light of Qatar National Vision (QNV) 2030. The survey is anonymous and your input is very essential in achieving the project objectives.

Section I – Background Information:

This section asks for some background information about you and your organization.

1. What is your gender, and nationality?
   ○ Male ○ Female | ○ Qatari ○ Non-Qatari, pl. specify:............... | Organization:.................

2. What is your highest educational degree, which specialization, which university, and when did you obtain it?..
   ....

3. Which of the following best describes your position at work? [Pls. check all that apply].

   ☐ Manager/Supervisor ☐ HR Specialist ☐ Training Specialist ☐ Other(s): ....

4. Please, classify your organization in terms of the major industry sector. [Pls. check all that apply].

   ☐ Energy ☐ Services ☐ Construction ☐ ICT
   ☐ Oil & gas ☐ Trading ☐ Education ☐ Government/Military
   ☐ Consulting ☐ Media ☐ Health/Bio-Tech ☐ Other: ....

5. Approximately, how many staff (full-time) does your organization currently employ?

   ○ <20 ○ 20-50 ○ 51-100 ○ 100-250 ○ More than 250, please specify: ....

6. What is the scope of operations of your organization? [Pls. check all that apply].
   ☐ National ☐ Regional (GCC) ☐ Middle East ☐ International

1

Section II – Competencies & Skills of Engineering Graduates in Qatar:

This section asks for your views on the supply of suitably qualified engineering graduates direct from universities in Qatar

7. **The following are 22 sets of generic engineering graduate skills classified under 4 dimensions. Please indicate under** "*Importance*" **column how much you perceive these skills to be important currently and in future for engineers (in light of QNV 2030)** (1= strongly disagree; 2= disagree; 3= Neutral; 4= agree; 5=strongly agree), **and specify under the** "*Satisfaction*" **column how much you are satisfied with the efficiency of undergraduate engineering education in preparing students with these skills to join the engineering practice** (1= strongly disagree; 2= disagree; 3= Neutral; 4= agree; 5=strongly agree)

Importance Currently 1 2 3 4 5	Importance 2030 1 2 3 4 5		Satisfaction 1 2 3 4 5
		Dimensions I: CORE ENGINEERING KNOWLEDGE & PRACTICE	
O O O O O	O O O O O	**DISCIPLINARILY ENGINEERING FUNDAMENTALS (Depth)** "Demonstrate in-depth technical knowledge in the engineering field of specialty and know-how in specific engineering discipline and apply this knowledge effectively through practice – in balance with non-technical skills"	O O O O O
O O O O O	O O O O O	**INTERDISCIPLINARY ENGINEERING KNOWLEDGE (Breadth)** "Develop and build a comprehensive interdisciplinary knowledge from a specific filed through a scholarly approach that incorporating theory and practice"	O O O O O
O O O O O	O O O O O	**MATH, PHYSICS, & SCIENCE FUNDAMENTALS** "Demonstrate essential mathematical skills and use, quantify, and manipulate numbers and their application accurately in any contexts"	O O O O O
O O O O O	O O O O O	**PRACTICAL EXPERIENCE** "Apply knowledge and skills, required for effective professional engineering practices, into real-world settings"	O O O O O
O O O O O	O O O O O	**ICT EXPERIENCE** "Use information and communication technology applications interactively in locating, managing, modeling, analyzing, evaluating, and processing information from multiple sources to effectively accomplish a given task"	O O O O O
O O O O O	O O O O O	**MULTIDISCIPLINARY KNOWLEDGE (Outside the engineering fields, e.g. social, humanities, business, etc.)** "Develop and build a multidisciplinary knowledge from other engineering fields through a scholarly approach that incorporating theory and practice"	O O O O O
		Dimensions II: COGNITION & THINKING	
O O O O O	O O O O O	**LIFELONG LEARNING** "Ability to continuously acquire new knowledge and skills for self and professional development at all levels (life and career)"	O O O O O
O O O O O	O O O O O	**PROBLEM SOLVING** "Use engineering knowledge systematically to identify, analyze, formulate, solve, and evaluate complex and multidisciplinary problems – by applying cognitive skills (logical, critical, and creative thinking)"	O O O O O

2

Importance Currently 1 2 3 4 5	Importance 2030 1 2 3 4 5		Satisfaction 1 2 3 4 5
O O O O O	O O O O O	**DECISION MAKING** "Apply personal and professional judgment, take-risk and initiative, in effectively making strategic decision and managing risks, from a range of alternatives, based-on available information in response to ambiguous and complex situation"	O O O O O
O O O O O	O O O O O	**SYSTEMS THINKING** "Think systematically in holistic engineering approach by understanding how an entire system works and the relationship among systems' elements and other systems, and viewing a problem as a part of an overall system"	O O O O O
O O O O O	O O O O O	**CRITICAL THINKING** "Use critical, conceptual, reflective, and rational thinking in drawing and evidence-based assessing systematic conclusions and finding underlying relationships for solutions"	O O O O O
O O O O O	O O O O O	**INNOVATION** "Add values through introducing new novel ideas, methods, directions, opportunities, and solutions that meet new requirements, through more effective products, processes, services, and technologies, that are readily available to stakeholders"	O O O O O
O O O O O	O O O O O	**DESIGN** "Design engineering products to meet specified needs and develop optimal solutions for complex engineering problems"	O O O O O
		Dimensions III: PROFESSIONAL & INTERPERSONAL	
O O O O O	O O O O O	**PROFESSIONALISM** "Maintain high-level of social and professional behavior, demonstrate accountability, accept responsibility, and achieve excellence in work and everyday life"	O O O O O
O O O O O	O O O O O	**ETHICS** "Understand ethical norms and demonstrate ethical reasoning, behavior, and integrity of social responsibility and service in any context"	O O O O O
O O O O O	O O O O O	**ADAPTABILITY** "Demonstrate ingenuity in addressing challenges and flexibility and adaptability for actively managing rapid and diverse change successfully in workplace"	O O O O O
O O O O O	O O O O O	**COMMUNICATION** "Communicate effectively, in various ways (verbal, oral, written), with all stakeholders (Public, Engineers) across all boundaries (cultural, language)"	O O O O O
O O O O O	O O O O O	**TEAMWORK** "Work & cooperate efficiently in a diverse, multicultural, and interdisciplinary team"	O O O O O
O O O O O	O O O O O	**FOREIGN LANGUAGE(s)** "Demonstrate language proficiency, in foreign language(s) (in particular English for non-native speakers) at both level, professional and technical"	O O O O O

3

Importance Currently 1 2 3 4 5	Importance 2030 1 2 3 4 5		Satisfaction 1 2 3 4 5
		Dimensions IV: BUSINESS/ MANAGEMENT	
O O O O O	O O O O O	**MANAGEMENT** "Awareness of business literacy & practice in the context of engineering, manage organization & Resources efficiently, develop processes, plan projects, take-risk"	O O O O O
O O O O O	O O O O O	**LEADERSHIP** "Engaged, interact, lead, and influence people effectively toward accomplishing a common or shared goal and contributing positively to diverse communities"	O O O O O
O O O O O	O O O O O	**ENTREPRENEURSHIP** "Identify, capture, and develop an opportunity to create real business and social value through converting invention into successful innovation"	O O O O O
O O O O O	O O O O O	OTHER COMPETENCIES, please specify, ...	O O O O O

8. **Please rank top 5 engineering skills listed in question 7 above in order of importance (starting with 1 for the most important):**

Rank 1 .... Rank 2 .... Rank 3 .... Rank 4 .... Rank 5 ....

9. **Among the skills/competencies listed in question 7 above, which ones do you think are not adequately emphasized or should be more emphasized in undergraduate engineering education?**..
....

10. **To what extent do you agree or disagree that the following groups possess the following? (1= strongly disagree; 2= disagree; 3= Neutral; 4= agree; 5=strongly agree)**

	Male 1 2 3 4 5	Female 1 2 3 4 5	Qatari 1 2 3 4 5	Non- Qatari 1 2 3 4 5	Resident 1 2 3 4 5	Overseas 1 2 3 4 5	Qatar university 1 2 3 4 5	Other universities in Qatar 1 2 3 4 5
Generic engineering skills	O O O O O	O O O O O	O O O O O	O O O O O	O O O O O	O O O O O	O O O O O	O O O O O
GPA	O O O O O	O O O O O	O O O O O	O O O O O	O O O O O	O O O O O	O O O O O	O O O O O
Work preparedness	O O O O O	O O O O O	O O O O O	O O O O O	O O O O O	O O O O O	O O O O O	O O O O O
Motivation	O O O O O	O O O O O	O O O O O	O O O O O	O O O O O	O O O O O	O O O O O	O O O O O
Management skills	O O O O O	O O O O O	O O O O O	O O O O O	O O O O O	O O O O O	O O O O O	O O O O O
Planning & organizing skills	O O O O O	O O O O O	O O O O O	O O O O O	O O O O O	O O O O O	O O O O O	O O O O O
Leadership skills	O O O O O	O O O O O	O O O O O	O O O O O	O O O O O	O O O O O	O O O O O	O O O O O

Qatar University
College of Engineering
HR survey

11. **To which extent do you agree that it is important for the following groups to master the following Management and Business competencies for their professional development in your organization?** (1= strongly disagree; 2= disagree; 3= Neutral; 4= agree; 5=strongly agree)

	1	2	3	4	5	1	2	3	4	5	1	2	3	4	5	1	2	3	4	5
	Male					**Female**					**Qatari**					**Non-Qatari**				
Project management skills	O	O	O	O	O	O	O	O	O	O	O	O	O	O	O	O	O	O	O	O
Risk management skills	O	O	O	O	O	O	O	O	O	O	O	O	O	O	O	O	O	O	O	O
Planning & organizing skills	O	O	O	O	O	O	O	O	O	O	O	O	O	O	O	O	O	O	O	O
Leadership skills	O	O	O	O	O	O	O	O	O	O	O	O	O	O	O	O	O	O	O	O

12. **Do you think that competencies & skills demanded by Qatar industry would change from now till 2022 & from 2022 till 2030?**

By 2022:	**By 2030:**
O No O Yes	O No O Yes
Please provide further details for your response;	**Please provide further details for your response;**
....	....
....	....
....	....
....	....
....	....

Section III – Recruitment & Employment:

This section asks for information regarding recruitment and employment in your organization.

13. **Please indicate your organizations degree of interest in recruiting engineers**(1= no interest; 2= low interest; 3= Neutral ; 4= moderate interest; 5= considerable interest)

	1	2	3	4	5		1	2	3	4	5		1	2	3	4	5
Fresh graduate	O	O	O	O	O	**Qatari**	O	O	O	O	O	**Qatar university**	O	O	O	O	O
Experienced (5+ years)	O	O	O	O	O	**Non-Qatari**	O	O	O	O	O	**Other universities**	O	O	O	O	O
												Pls specify:..........					
Male	O	O	O	O	O	**Resident**	O	O	O	O	O	**BSc**	O	O	O	O	O
Female	O	O	O	O	O	**Overseas**	O	O	O	O	O	**MSc**	O	O	O	O	O
												PhD	O	O	O	O	O

14. **How many engineers does your organization currently employ from the categories below on full-time basis?**

Male Qatari Engineers: .... **Female Qatari Engineers:**............................

Male non-Qatari Engineers: .... **Female non-Qatari Engineers:** ....

15. **How much is the average projected increase/decrease in employing new engineering staff PER-YEAR over the next three to five years?**

O Increase (Per-Year%): ....%
O Decrease (Per-Year%): ....%
O No Increase or Decrease (Same as current)
O NA (No Idea or Projection is available)

Qatar University
College of Engineering
HR survey

16. Do you recruit engineering graduates who are not Qatar residents?

○ No ○ Yes

If yes; what is the reason? [Pls. check all that apply]

☐They are hardworking and has better performance
☐They are willing to work for less
☐Recruiting the very best talent available
☐Their disciplinary engineering field is not found in Qatar
☐Other: ....
☐Insufficient candidates with the right skills in Qatar
☐More experienced than graduates from local universities
☐They have the needed skills
☐They have better work professionalism

17. What are the engineering disciplines that fill in your organization **now**? & what are the disciplines you expect to recruit to satisfy the needs and the plans of your company over the **next** 5-15 years (till 2022 & 2030)? And what is the **estimated percentage** of each discipline? [Pls. check all that apply]

Engineering Discipline	Now	Estimated % now	Till 2022	Estimated % till 2022	2022-2030	Estimated % 2022-2030
Architecture Engineering	☐	....%	☐	....%	☐	....%
Biomedical Engineering	☐	....%	☐	....%	☐	....%
Civil Engineering	☐	....%	☐	....%	☐	....%
Chemical Engineering	☐	....%	☐	....%	☐	....%
Computer/ software Engineering	☐	....%	☐	....%	☐	....%
Electrical Engineering	☐	....%	☐	....%	☐	....%
Automation/Control	☐	....%	☐	....%	☐	....%
Industrial & systems Engineering	☐	....%	☐	....%	☐	....%
Mechanical Engineering	☐	....%	☐	....%	☐	....%
Petroleum Engineering	☐	....%	☐	....%	☐	....%
Other(s):	☐	....%	☐	....%	☐	....%

Section IV – Industry- Academia Collaboration:

This section asks about your organization's practices regarding the linkages between Industry and Academia in Qatar.

18. Does your organization have links with engineering universities/colleges?

○ No ○ Yes

If yes; for each of the following means of collaboration, please indicate through which means your company has collaborated and/or prefers to collaborate with universities. [Pls. check all that apply]

Means of Collaboration	Already collaborating				Would like to collaborate		
	Yes	No	If yes mention university name		Yes	No	If yes mention university name
Sponsorships for students	☐	☐	....		☐	☐	....
Engineering and career events	☐	☐	....		☐	☐	....
Graduate recruitment	☐	☐	....		☐	☐	....
Offer internships to graduates	☐	☐	....		☐	☐	....
Provide 'real-life' projects and resources	☐	☐	....		☐	☐	....
University provides off-site workforce training for our employee	☐	☐	....		☐	☐	....
Partner with universities for research and innovation	☐	☐	....		☐	☐	....
Others	☐	☐	....		☐	☐	....

19. How much does your organization perceive collaboration with universities to be beneficial/ important for both sides?

○ Very important ○ Important ○ Not important ○ Not important at all ○ N/A

Qatar University
College of Engineering
HR survey

20. Do you think there are any barriers that limit interaction with universities?
   O No   O Yes

If yes; please indicate to which extent you agree with the following factors being a reason in limiting/preventing your organization from interacting with Universities. [pls. check all that applies]

Inadequate infrastructure	☐	There is no value or benefit to our company from such collaboration	☐
Inadequate laboratory facilities	☐	We had a bad experience with university in the past	☐
Lack of autonomy to work with academia	☐	The process is too complicated	☐
The university structure is not adapted to the needs of industrial collaborations	☐	Local universities are not interested/ no collaboration from university side	☐
The collaboration is time consuming	☐	Difficulty fitting in with university timetable	☐
Lack of organizational/ structural office/ unit to coordination	☐	Other(s):	☐

Section V – Comments & Further Suggestions:

21. Please, add any further comments and/or recommendations you may consider to be important in helping local Engineering Colleges to prepare skilled engineers capable of meeting industry needs and governmental strategic plans in Qatar to fulfill the QNV 2030.

....
....
....
....
....

Thanks for your participation in this survey…

Industrial Training Sector Survey

Qatar University
College of Engineering
Training Department survey

This survey is conducted as part of large-scale study that focuses on preparedness of Engineering Graduates in Qatar currently and in light of Qatar National Vision (QNV) 2030. The survey is anonymous and your input is very essential in achieving the project objectives.

Section I – Background Information:

This section asks for some background information about you and your organization

1 **What is your gender, and nationality?**
O Male O Female | O Qatari O Non-Qatari, pls. specify: .... | Organization: ....

2 **What is your highest educational degree, which specialization, which university, and when did you obtain it?**..
....

3 **What is your position at work?** ....

4 **Which of the following groups do you offer training for? [pls check all that apply]**
☐ Engineering Students ☐ Newly hired engineering graduates ☐ Practicing engineers ☐ Others:................

Section II – Competencies & Skills of Engineering Graduates in Qatar:

This section asks for your views on the supply of suitably qualified engineering graduates direct from universities in Qatar

5 **The following are 22 sets of generic engineering graduate skills classified under 4 dimensions. Please indicate under** "*Importance*" **column how much you perceive these skills to be important currently and in future for engineers (in light of QNV 2030)** (1= strongly disagree; 2= disagree; 3= Neutral; 4= agree; 5=strongly agree), **and specify under the** "*Satisfaction*" **column how much you are satisfied with the efficiency of undergraduate engineering education in preparing students with these skills to join the engineering practice** (1= strongly disagree; 2= disagree; 3= Neutral; 4= agree; 5=strongly agree)

Importance Currently	Importance 2030		Satisfaction
1 2 3 4 5	1 2 3 4 5		1 2 3 4 5
		Dimensions I: CORE ENGINEERING KNOWLEDGE & PRACTICE	
O O O O O	O O O O O	**DISCIPLINARILY ENGINEERING FUNDAMENTALS (Depth)** "Demonstrate in-depth technical knowledge in the engineering field of specialty and know-how in specific engineering discipline and apply this knowledge effectively through practice – in balance with non-technical skills"	O O O O O
O O O O O	O O O O O	**INTERDISCIPLINARY ENGINEERING KNOWLEDGE (Breadth)** "Develop and build a comprehensive interdisciplinary knowledge from a specific filed through a scholarly approach that incorporating theory and practice"	O O O O O
O O O O O	O O O O O	**MATH, PHYSICS, & SCIENCE FUNDAMENTALS** "Demonstrate essential mathematical skills and use, quantify, and manipulate numbers and their application accurately in any contexts"	O O O O O
O O O O O	O O O O O	**PRACTICAL EXPERIENCE** "Apply knowledge and skills, required for effective professional engineering practices, into real-world settings"	O O O O O
O O O O O	O O O O O	**ICT EXPERIENCE** "Use information and communication technology applications interactively in locating, managing, modeling, analyzing, evaluating, and processing information from multiple sources to effectively accomplish a given task"	O O O O O
O O O O O	O O O O O	**MULTIDISCIPLINARY KNOWLEDGE (Outside the engineering fields, e.g. social, humanities, business, etc.)** "Develop and build a multidisciplinary knowledge from other engineering fields through a scholarly approach that incorporating theory and practice"	O O O O O
		Dimensions II: COGNITION & THINKING	
O O O O O	O O O O O	**LIFELONG LEARNING** "Ability to continuously acquire new knowledge and skills for self and professional development at all levels (life and career)"	O O O O O
O O O O O	O O O O O	**PROBLEM SOLVING** "Use engineering knowledge systematically to identify, analyze, formulate, solve, and evaluate complex and multidisciplinary problems – by applying cognitive skills (logical, critical, and creative thinking)"	O O O O O

2

Importance Currently 1 2 3 4 5	Importance 2030 1 2 3 4 5		Satisfaction 1 2 3 4 5
O O O O O	O O O O O	**DECISION MAKING** "Apply personal and professional judgment, take-risk and initiative, in effectively making strategic decision and managing risks, from a range of alternatives, based-on available information in response to ambiguous and complex situation"	O O O O O
O O O O O	O O O O O	**SYSTEMS THINKING** "Think systematically in holistic engineering approach by understanding how an entire system works and the relationship among systems' elements and other systems, and viewing a problem as a part of an overall system"	O O O O O
O O O O O	O O O O O	**CRITICAL THINKING** "Use critical, conceptual, reflective, and rational thinking in drawing and evidence-based assessing systematic conclusions and finding underlying relationships for solutions"	O O O O O
O O O O O	O O O O O	**INNOVATION** "Add values through introducing new novel ideas, methods, directions, opportunities, and solutions that meet new requirements, through more effective products, processes, services, and technologies, that are readily available to stakeholders"	O O O O O
O O O O O	O O O O O	**DESIGN** "Design engineering products to meet specified needs and develop optimal solutions for complex engineering problems"	O O O O O
		Dimensions III: PROFESSIONAL & INTERPERSONAL	
O O O O O	O O O O O	**PROFESSIONALISM** "Maintain high-level of social and professional behavior, demonstrate accountability, accept responsibility, and achieve excellence in work and everyday life"	O O O O O
O O O O O	O O O O O	**ETHICS** "Understand ethical norms and demonstrate ethical reasoning, behavior, and integrity of social responsibility and service in any context"	O O O O O
O O O O O	O O O O O	**ADAPTABILITY** "Demonstrate ingenuity in addressing challenges and flexibility and adaptability for actively managing rapid and diverse change successfully in workplace"	O O O O O
O O O O O	O O O O O	**COMMUNICATION** "Communicate effectively, in various ways (verbal, oral, written), with all stakeholders (Public, Engineers) across all boundaries (cultural, language)"	O O O O O
O O O O O	O O O O O	**TEAMWORK** "Work & cooperate efficiently in a diverse, multicultural, and interdisciplinary team"	O O O O O
O O O O O	O O O O O	**FOREIGN LANGUAGE(s)** "Demonstrate language proficiency, in foreign language(s) (in particular English for non-native speakers) at both level, professional and technical"	O O O O O

3

Importance Currently 1 2 3 4 5	Importance 2030 1 2 3 4 5		Satisfaction 1 2 3 4 5
		Dimensions IV: BUSINESS/ MANAGEMENT	
O O O O O	O O O O O	**MANAGEMENT** "Awareness of business literacy & practice in the context of engineering, manage organization & Resources efficiently, develop processes, plan projects, take-risk"	O O O O O
O O O O O	O O O O O	**LEADERSHIP** "Engaged, interact, lead, and influence people effectively toward accomplishing a common or shared goal and contributing positively to diverse communities"	O O O O O
O O O O O	O O O O O	**ENTREPRENEURSHIP** "Identify, capture, and develop an opportunity to create real business and social value through converting invention into successful innovation"	O O O O O
O O O O O	O O O O O	OTHER COMPETENCIES, please specify, ...	O O O O O

6. Please rank top 5 engineering skills listed in question 5 above in order of importance (starting with 1 for the most important):

Rank 1 .... Rank 2 .... Rank 3 .... Rank 4 .... Rank 5 ....

7. Among the skills/competencies listed in question 5 above, which ones do you think are not adequately emphasized or should be more emphasized in undergraduate engineering education?..

....

8. To what extent do you agree or disagree that the following groups possess the following? (1= strongly disagree; 2= disagree; 3= Neutral; 4= agree; 5=strongly agree)

	Male 1 2 3 4 5	Female 1 2 3 4 5	Qatari 1 2 3 4 5	Non- Qatari 1 2 3 4 5	Resident 1 2 3 4 5	Overseas 1 2 3 4 5	Qatar university 1 2 3 4 5	Other universities in Qatar 1 2 3 4 5
Generic engineering skills	O O O O O	O O O O O	O O O O O	O O O O O	O O O O O	O O O O O	O O O O O	O O O O O
GPA	O O O O O	O O O O O	O O O O O	O O O O O	O O O O O	O O O O O	O O O O O	O O O O O
Work preparedness	O O O O O	O O O O O	O O O O O	O O O O O	O O O O O	O O O O O	O O O O O	O O O O O
Motivation	O O O O O	O O O O O	O O O O O	O O O O O	O O O O O	O O O O O	O O O O O	O O O O O
Management skills	O O O O O	O O O O O	O O O O O	O O O O O	O O O O O	O O O O O	O O O O O	O O O O O
Planning & organizing skills	O O O O O	O O O O O	O O O O O	O O O O O	O O O O O	O O O O O	O O O O O	O O O O O
Leadership skills	O O O O O	O O O O O	O O O O O	O O O O O	O O O O O	O O O O O	O O O O O	O O O O O

4

Qatar University
College of Engineering
Training Department survey

9. To which extent do you agree that it is **important** for the following groups to **master** the following **Management and Business competencies** for their professional development in your organization? (1= strongly disagree; 2= disagree; 3= Neutral; 4= agree; 5=strongly agree)

	1	2	3	4	5	1	2	3	4	5		1	2	3	4	5	1	2	3	4	5
	Male					**Female**						**Qatari**					**Non-Qatari**				
Project management skills	O	O	O	O	O	O	O	O	O	O		O	O	O	O	O	O	O	O	O	O
Risk management skills	O	O	O	O	O	O	O	O	O	O		O	O	O	O	O	O	O	O	O	O
Planning & organizing skills	O	O	O	O	O	O	O	O	O	O		O	O	O	O	O	O	O	O	O	O
Leadership skills	O	O	O	O	O	O	O	O	O	O		O	O	O	O	O	O	O	O	O	O

10. Do you think that competencies & skills demanded by Qatar industry would change from now till 2022 & from 2022 till 2030?

By 2022:	By 2030:
O No O Yes	O No O Yes
Please provide further details for your response;	**Please provide further details for your response;**
....	....
....	....
....	....
....	....
....	....

Section III – Training & Development of employees:

This section asks for information regarding training and development programs.

11. **Please indicate your organizations degree of interest in training engineers;** (1= no interest; 2= low interest; 3= Neutral ; 4= moderate interest; 5= considerable interest)

	1	2	3	4	5		1	2	3	4	5		1	2	3	4	5
Fresh graduate	O	O	O	O	O	**Qatari**	O	O	O	O	O	**Qatar university**	O	O	O	O	O
Experienced (5+ years)	O	O	O	O	O	**Non-Qatari**	O	O	O	O	O	**Other universities**	O	O	O	O	O
												Pls specify:..........					
Male	O	O	O	O	O	**Resident**	O	O	O	O	O	**BSc**	O	O	O	O	O
Female	O	O	O	O	O	**Overseas**	O	O	O	O	O	**MSc**	O	O	O	O	O
												PhD	O	O	O	O	O

12. Does your organization offer **continuous development and training** for **fresh** (newly hired) **engineers**?
O No O Yes
If yes, what kind of training?..
....
....

13. Does your organization offer continuous development and training for **practicing engineers**?
O No O Yes
If yes, what kind of training?..
....
....

5

Qatar University
College of Engineering
Training Department survey

14. Do you think there are any barriers from training?

O No O Yes

If yes; Please indicate to which extent you agree with the following factors being a reason in limiting/preventing your organization from training engineers. (1= strongly disagree; 2= disagree; 3= Neutral; 4= agree; 5=strongly agree)

	1	2	3	4	5
Experience in organizations is recent	O	O	O	O	O
Training department is not well structured	O	O	O	O	O
The perception of training is administrative; focuses on booking flights, courses, ...	O	O	O	O	O
Small numbers of trainees that sometimes don't merit certain approaches where economy of scale is needed	O	O	O	O	O
Organization is small	O	O	O	O	O
Other:	O	O	O	O	O
Other:	O	O	O	O	O

Section IV – Industry- Academia Collaboration:

This section asks about your organization's practices regarding the linkages between Industry and Academia in Qatar.

15. Does your organization have links with engineering universities/colleges?

O No O Yes

If yes; for each of the following means of collaboration, please indicate through which means your company has collaborated and/or prefers to collaborate with universities. [Pls. check all that apply]

Means of Collaboration	Already collaborating				Would like to collaborate		
	Yes	No	If yes mention university name		Yes	No	If yes mention university name
Engineering and career events	☐	☐	....		☐	☐	....
Offer internships to graduates	☐	☐	....		☐	☐	....
Provide 'real-life' projects and resources	☐	☐	....		☐	☐	....
University provides off-site workforce training for our employee	☐	☐	....		☐	☐	....
Others	☐	☐	....		☐	☐	....

16. How much does your organization perceive collaboration with universities to be beneficial/ important for both sides?

O Very important O Important O Not important O Not important at all O N/A

17. Do you think there are any barriers that limit interaction with universities?

O No O Yes

If yes; please indicate to which extent you agree with the following factors being a reason in limiting/preventing your organization from interacting with Universities. [pls. check all that applies]

Inadequate infrastructure	☐	There is no value or benefit to our company from such collaboration	☐
Lack of autonomy to work with academia	☐	We had a bad experience with university in the past	☐
The university structure is not adapted to the needs of industrial collaborations in training	☐	The process is too complicated	☐
The collaboration is time consuming	☐	Local universities are not interested	☐
Lack of organizational/ structural office/ unit to coordination	☐	Other(s):	☐

References

2028 Vision for Mechanical Engineering: A report of the Global Summit on the Future of Mechanical Engineering. (2008). *Notes*. New York, USA. Retrieved from www.asme.org

Abdullah, S., Zaharim, A., Harris, S. M., Omar, M. Z., Basri, H., & Mohamed, N. A. N. I. K. (2007). Engineering education: Using technical attributes to analyse the employers' expectation of future engineering graduates in Malaysia. *Engineering Education*, 191–195.

ACER. (2002). *Employability skills for australian industry: Literature review and framework development*.

Agogino, A. M. (n.d.). *Educating the engineer of 2020: Global visions for the New Century NAE engineer of 2020*.

Allan, M., & Chisholm, C. U. (2009). The formation of the engineer for the 21st century—A global perspective. In *20th Australasian Association for Engineering Education Conference* (pp. 447–452).

Anderson, D., Somerville, M., Berbeco, H., Bourne, J. R., Crisman, J., Dabby, D., et al. (2005). The olin curriculum: Thinking toward the future. *IEEE Transactions on Education, 48*(1), 198–205.

APEC. (2000). *Towards knowledge based economies in APEC*. APEC Secretariat.

ASCE. (2007). *The vision for civil engineering in 2025*. Virginia. Retrieved from www.asce.org

ASME. (n.d.). *The state of mechanical engineering: Today and beyond*.

Barrie, S. C. (2007, August). A conceptual framework for the teaching and learning of generic graduate attributes. *Studies in Higher Education, 32*(4), 439–458.

Boston, M., & Boston, M. B. (n.d.). Information technology workforce skills.

Bourn, D., & Neal, I. (2008). The global engineer: Incorporating global skills within UK higher education of engineers. *Engineer*.

Bowman, K. (2010, March). Background paper for the AQF Council on generic skills.

Calfee, C. (2009). Galloway' s 21st century engineer : An essay review Robert Calfee University of California, Riverside Stanford University Thomas Stahovich University of California, Riverside Galloway. Patricia, D. (2008) The 21st century engineer: A proposal for E, *12*(14).

Casner-Lotto, J., & Barrington, L. (2006). *Are they really ready to work? Employers' perspectives on the basic knowledge and applied skills of new entrants to the 21st century US workforce*. Retrieved from http://www.eric.ed.gov/ERICWebPortal/recordDetail?accno=ED519465

CBI. (2007). *Time well spent Embedding employability in work experience*. London, UK.

CBI. (2009). *Jobs for the future: The business vision for sustainable employment in the UK*. UK.

CBI. (2012). *Learning to grow: What employers need from education and skills—Education and skills survey 2012*. London, UK. Retrieved from www.cbi.org.uk

Chan, A. D. C., & Fishbein, J. (2009). Aglobal engineer for the global community. *The Journal of Policy Engagement, 1*(2), 4–9.

Chang, Y., Atkinson, D., & Hirleman, E. D. (2009). International research and engineering education: Impacts and best practices education: Impact and best practices. *Online Journal for Global Engineering Education, 4*(2).

M. Abdulwahed and M.O. Hasna, *Engineering and Technology Talent for Innovation and Knowledge-Based Economies*,
DOI 10.1007/978-3-319-46439-8

Childs, P., & Gibson, P. (2010). Graduating professional engineers and management skills-are they adequate for the workplace? *3rd International Symposium for Engineering Education, 2010, University College Cork, Ireland*, 1–9. Retrieved from http://ro.uow.edu.au/chsd/34/

Clausen, J. M. (2004). *The danish strategy, Denmark's opportunities in the global knowledge society.* On behalf of Strategy Group.

Conference Board of Canada. (2000). *Employability skills 2000+*. Canada.

Continental, A. G. (2006). *Final report of the global engineering excellence initiative; educating the next generation of engineers for the global workplace* (p. 114). Hanover/Germany. ISBN: 3-9811322-1-1

Co-sponsors, W., Committee, S., Lauby, M., Liu, C., Niebur, D., Foundation, N. S., et al. (2008). National science foundation workshop on the future power engineering workforce. *Systems Engineering*.

Criteria, J. C., Programs, P. E., & Kaikan, K. (2012). JABEE common criteria for accreditation of professional education programs.

Danielson, S. (2011). ASME vision 2030: Helping to inform mechanical engineering education. In *41st ASEE/IEEE frontiers in education conference, 12–15, 2011* (pp. 1–6). Rapid City, SD. Retrieved from http://ieeexplore.ieee.org/xpls/abs_all.jsp?arnumber=6143065

David, P. A., & Foray, D. (2002). An introduction to the economy of the knowledge society. *International Social Science Journal, 54*(171), 9–23.

Davis, H., Evans, T., & Hickey, C. A. (2006), Knowledge based economy landscape: Implications for tertiary education and research training in Australia. *Journal of Higher Education Policy and Management, 28*(3), 231–244.

Dawe, S. (2002). *Focussing on generic skills in training packages*. Australia. Retrieved from http://citeseerx.ist.psu.edu/viewdoc/download?doi=10.1.1.198.3625&rep=rep1&type=pdf

DEEWR. (2012). *Employability skills framework* (pp. 1–24). Retrieved from http://deewr.gov.au/

DEST. (2006). *Employability skills from framework to practice: An introductory guide for trainers and assessors*. Melbourne.

DEST. (2007). *Graduate employability skills: Prepared for the business, industry and higher education collaboration council*. Melbourne. ISBN: 0 642 77701 2

Dodridge, M. (2001). Session 8D2 developing occupational standards in engineering degree programmes session 8D2 (pp. 23–28).

Duderstadt, B. J. J. (2011). *Heartland papers a master plan for higher education in the midwest*. Chicago, Illinois, USA.

Duderstadt, J. J. (2008a). *Engineering for a changing world a roadmap to the future of a roadmap to the future of. engineering*.

Duderstadt, J. J. (2008b). Engineering for a changing world a roadmap to the future of american engineering practice, research, and education (Short Paper). *Engineering Education*.

Duderstadt, J. J. (2010). *Holistic engineering education* (D. Grasso & M. B. Burkins, Eds.), (pp. 17–35). doi:10.1007/978-1-4419-1393-7

Dudman, A., & Wearne, S. (2003, March). Professional engineers' needs for managerial skills and expertise. Centre for Research in the Management of Projects University of Manchester Institute of Science & Technology. Retrieved from http://www.engc.org.uk/ecukdocuments/internet/documentlibrary/ProfessionalEngineersneedManagementSkills.pdf

Dutta, S. & Mia, I. (2011) *The global information technology report 2010-2011*. Transformations 2.0—(10th Anniversary Edition). From: INSEAD and World Economic Forum.

Economic Importance of Engineers (2009). Prepared by DKM Economic Consultants Ltd. For Engineers Ireland.

Edinburgh. (2012). *Edinburgh Napier university conscious employability model*. Edinburgh. Retrieved from http://www.napier.ac.uk/oldcontent/standoutfromthecrowd/Pages/ConsciousEmployabilityModel.aspx

Embracing Scientific and Technological Change. *Shanghai Manual—A Guide for Sustainable Urban Development in the 21st Century*, Chapter 7.

Engineering Accreditation Accreditation. (2004). (65).

ERT. (1995). *Education for Europeans—Towards the learning society*. Brussels: The European Round Table of Industrialists.

Felder, R. M. (1998). ABET criteria 2000: An exercise in engineering problem solving. *Chemical Engineering Education, 32*(2), 126–127.

Finegold, D., & Notabartolo, A. (2010). 21st-century competencies and their impact: an interdisciplinary literature review. *Research on 21st Century Competencies, National Research Council*, 1–50. Retrieved from http://onlinelibrary.wiley.com/doi/10.1002/cbdv.200490137/abstract

Fung, M. C. D., Lee, W. Y. W., & Wong, S. L. P. (2006). A new measure of generic competencies.

Gibb, J. (2004). Generic skills in vocational education and training: Research readings. *National Centre for Vocational Education Research*. Retrieved from http://www.ncver.edu.au

Goel, S. (2006). Competency focused engineering education with reference to IT related disciplines : Is the Indian system ready for transformation? *Journal of Information Technology Education, 5*.

Hager, P., & Holland, S. (2006). *Graduate attributes, learning and employability* (pp. 1–15). Springer. http://www.springer.com/education+%26+language/book/978-1-4020-5341-2.

Holsti, O. R. (1969). *Content analysis for the social sciences and humanities*. Reading, MA: Addison-Wesley.

Hounsell, D. (2011). *Graduates for the 21st century: Integrating the enhancement themes—Institutional activities*. Edinburgh. Retrieved from www.enhancementthemes.ac.uk

HRA. (2013). *IT takes more than a major: Employer priorities for college learning and student success an online survey among employers conducted on behalf of: The Association Of American Colleges And Universities* (pp. 1–15). Washington, D.C., USA.

Hundley, S. (2012). AC 2012-4233: Attributes of a global engineer: Field-informed perspectives, recommendations, and implications. Retrieved from http://www.asee.org/public/conferences/8/papers/4233/view

Hundley, S., Fox, P., Brown, L. G., Jacobs, A., Didion, C., Sayre, D. R., et al. (2012). *Attributes of a global engineer: Field-informed perspectives, recommendations, and implications*. American Society for Engineering Education, no. AC 2012–4233.

IEA. (2009). *Graduate attributes and professional competency* (pp. 1–15). Retrieved from http://www.washingtonaccord.org/GradProfiles.cfm

IFEES. (n.d.). The attributes of a global engineer project. In *International federation of attributes of a global engineer project focus group event at WEEF, engineering education societies* (pp. 1–10). Buenos Aires, Argentina.

Jamieson, L., Brophy, S., House, N., Harris, M., Delaurentis, D., Howell, K., et al. (2009) Purdue's engineer of 2020: The journey. In *2009 Annual Conference & Exposition. New Learning Paradigms II*.

Kastenberg, W., Hauser-Kastenberg, G., & Norris, D. (2006). An approach to undergraduate engineering education for the 21st century. *Proceedings of the Frontiers in Education. 36th Annual Conference* (pp. 23–28). doi:10.1109/FIE.2006.322502

Kerr, I. R. (2010). Futures thinking for engineering and engineers Australia's continuing professional development process. *Australasian Journal of Engineering Education, Institution of Engineers Australia, 16*(1), 13–20.

King, R. (2008). *Engineers for the future: Addressing the supply and quality of Australian engineering graduates for the 21st century* (pp. 1–137). ISBN: 978-0-9805211-0-8

Kirkpatrick, A. T. (2011). AC 2011-2804: Vision 2030 creating the future of mechanical engineering education.

Klein-gardner, S. S. (n.d.). Defining global competence for engineering students. In *118th ASEE 2011 Annual conference & exposition*. B.C. Canada.

Knight, D. B. (2012). In search of the engineers of 2020: An outcome-based typology of engineering undergraduates, AC 2012-3337. In *Proceedings of the 119th Annual Conference of the American Society for Engineering Education*. San Antonio, TX.

Kolbe, R. H., & Burnett, M. S. (1991). Content-analysis of an for examination improving applications with directives and reliability objectivity. *The Journal of Consumer Research, 18* (2), 243–250.

Li, T., & Cavusgil, S. (1995). A classification and assessment of research streams in international marketing. *International Business Review, 4*(3), 251–277.

Llanes, C. C. (2008). Outcomes-based education: The TIP experience. *Technological Institute of the Philippines*, 1–7.

Male, S. A., Bush, M. B., & Chapman, E. S. (2011). Understanding generic engineering competencies, *17*(3), 147–156.

Male, S. A. (2009). Identification of competencies required by engineers graduating in Australia. In *Proceedings of the 2009 AaeE Conference* (pp. 1–6). Melbourne, Australia. Retrieved from http://www.education.uwa.edu.au/__data/assets/pdf_file/0009/1898181/Identification-of-competencies-required-by-engineers-graduating-in-Australia.pdf

Male, S. A. (2010a). Perceptions of competency deficiencies in engineering graduates. *Australasian Journal of Engineering Education, 16*(1), 55–68. Retrieved from http://www.engineersmedia.com.au/journals/aaee/pdf/AJEE_16_1_MaleF2.pdf

Male, S. A. (2010b). Generic engineering competencies: A review and modelling approach. *Education Research and Perspectives, 37*(1), 25–51. Retrieved from http://erpjournal.net/wp-content/uploads/2012/07/ERPV37-1_Male-S.-2010.-Generic-Engineering-Competencies.pdf

Male, S., & Chapman, E. (2005). Assessing the generic competencies of engineering graduates: Preliminary report from an ongoing research program. In *Proceedings of the 2005 ASEE/AaeE 4th Global Colloquium on Engineering Education.* Australasian Association for Engineering Education. Retrieved from http://www.oii.uwa.edu.au/__data/assets/pdf_file/0012/1898229/Assessing-the-generic-competencies-of-engineering-graduates.pdf

Markes, I. (2006). A review of literature on employability skill needs in engineering. *European Journal of Engineering Education, 31*(6), 637–650. doi:10.1080/03043790600911704.

Martin, R., Villeneuve-smith, F., Marshall, L., & Mckenzie, E. (2008). *Employability skills explored. ...: Learning and skills ....* London. Retrieved from http://norfolkunites.org.uk/media/pdf/ese.pdf

Mena, I. B., Zappe, S. E., & Litzinger, T. A. (2012). Preparing the engineer of 2020: Analysis of alumnfi data. In *2012 ASEE Annual Conference, American Society for Engineering Education.*

Michigan. (2009). *Michigan engineering 2020, the commission on undergraduate engineering education: Curriculum for the 21st century.* Rivers. Retrieved from http://www.engin.umich.edu/admin/adue/undergradcommission/

Mishra, S. (2010). Engineering curricula in the 21st century: The global scenario and challenges for india. *Journal of Engineering, Science and Management Education, 1*(1), 29–33.

MSA. (2012). *Foundation skills in the manufacturing industry* (pp. 1–20).

NAE. (2004). *The engineer of 2020: Visions of engineering in the New Century.*

NAE. (2005). Educating the engineer of 2020: Adapting engineering education to the new century. *IEEE Engineering Management Review*, 1–208. doi:10.1109/EMR.2009.4804343

NAE. (2011). *Assessing 21st century skills.* Washington, DC 20001.

NCVER. (2003). *Defining generic skills at a glance* (pp. 1–12).

NESP. (2010). *Evaluation report: National employability skills program.* Canada. Retrieved from http://scholar.google.com/scholar?hl=en&btnG=Search&q=intitle:Evaluation+Report:+National+Employability+Skills+Program#1

Nguyen, D. (1998). The essential skills and attributes of an engineer: a comparative study of academics, industry personnel and engineering students. *Global Journal of Engineering Education*, 2(1), 65–76. Retrieved from http://citeseerx.ist.psu.edu/viewdoc/download?doi=10.1.1.124.1502&rep=rep1&type=pdf

Nguyen, N. D., Yoshinari, Y., & Shigeji, M. (2005). University education and employment in Japan: Students' perceptions on employment attributes and implications for university education. *Quality Assurance in Education, 13*(3), 202–218. doi:10.1108/09684880510607945

NHEAP. (2007). *Malaysia national higher education action plan 2007-2010.*

Noakes, N. (2004). *Employability, lifelong learning, personal development planning and eportfolios at HKUST* (pp. 1–8). Retrieved from http://repository.ust.hk/dspace/handle/1783.1/1710

Nor, N. M., Rajab, N., & Ismail, K. M. (2008). Educating the engineer of 2020: Malaysian scenario. In *International Conference on Engineering Education: New Challenges in Engineering Education and Research in the 21st Century*. Budapest, Hungary.

NRC. (2007). *Rising above the gathering storm: Energizing and employing America for a brighter economic future committee on prospering in the global economy of the 21st century: An agenda for American science and technology*. Washington, DC: National Academy Press.

NSB. (2007). *Moving forward to improve engineering education* (p. 58). National Science Board. NSB-07-122

Obama's campaign to improve the participation and performance of America's students in science, technology, engineering, and mathematics (STEM). Educate to Innovate. (2009). http://www.whitehouse.gov/issues/education/educate-innovate (Accessed June 26, 2011).

OECD. (1996). *The knowledge-based economy*. Paris: Organization for Economic Co-Operation and Development.

OECD. (2002). *Definition and Selection of competencies (DeSeCo) theoretical and conceptual foundations: Strategy paper*. Organisation for Economic Co-operation and Development.

Osman, W. D. (2011). *Higher education in Egypt and needed employability skills in the domestic labor market: Case study from AIN Shams University, Faculty of engineering*. The American University in Cairo.

P21. (2008). *21st Century skills, education & competitiveness: A resource and policy guide*. Tucson, USA. Retrieved from http://www.pkwy.k12.mo.us/projectParkway/File/21st_century_skills_education_and_competitiveness_guide.pdf

Palmer, B., Terenzini, P. T., McKenna, A. F., Harper, B. J., & Merson, D. (2011). Design in context: Where do the engineers of 2020 learn this skill. In *Proceedings of the 118th Annual Conference of the American Society for Engineering Education* (Vol. 2020). Vancouver, Canada. Retrieved from http://146.186.208.241:8082/Plone/e2020/asee-conference-presentations/asee2011-contextual-competence-final-paper.pdf

Parkinson, A. (2009). The rationale for developing global competence competence. *Online Journal for Global Engineering Education*, *4*(2). Retrieved from http://digitalcommons.uri.edu/ojgee/vol4/iss2/2

Patil, A., & Codner, G. (2008). Global accreditation for the global engineering attributes: A way forward. In *Proceedings of the 2008 AaeE Conference, Yeppoon* (pp. 1–8).

Pinelli, T. E., & Haynie, W. J. (2010, Spring). A case for the nationwide inclusion of engineering in the K-12 curriculum via technology education. *Journal of Technology Education*, *21*(2), 52–68.

Pomales-García, C., & Liu, Y. (2007). Excellence in engineering education: Views of undergraduate engineering students. *Journal of Engineering Education*, 253–262.

Pritchard, J. (2013). *The importance of soft skills in entry-level employment and postsecondary student success: Perspectives from employers and community colleges*. Seattle, USA. Retrieved from http://www.seattlejobsinitiative.com/wp-content/uploads/SJI_SoftSkillsReport_vFINAL_1.17.13.pdf

Programme, E., & Manual, A. (2003). *Engineering accreditation council Malaysia* (pp. 1–20).

PyeTait. (2011). *Ready for Work? The work readiness of entrants to the UK Construction sector who have work-related qualifications* (pp. 1–72). Harrogate, UK. Retrieved from www.pyetait.com

Qatar National Development Strategy 2011-2016 (2011)

Rabl, M., & Hillmer, G. (2012). The cultivation of engineering talent. In *SEFI 40TH Annual Conference*. Thessaloniki, Greece. Retrieved from http://www.sefi.be/conference-2012/Papers/Abstracts/040sum.pdf

RAE. (2007). *Educating engineers for the 21st century educating engineers for the 21st century*. London, UK: The Royal Academy of Engineering.

Rajala, S. A. (2012). Beyond 2020: Preparing engineers for the future. *Proceedings of the IEEE, 100* (Special Centennial Issue), (pp. 1376–1383). doi:10.1109/JPROC.2012.2190169

Ray, T., & Waterhouse, P. (2008). *Whose responsibility? Literature review & methodology—Support document.*

REFLEX. (2007). *The flexible professional in the knowledge society: General results of the REFLEX project.* Netherlands. Retrieved from http://ideas.repec.org/p/ner/maastr/urnnbnnlui27-20754.html

Saad, M., Robani, A., Jano, Z., & Majid, I. (2013). Employers' perception on engineering, information and communication technology (ICT) students' employability skills. *Journal of Engineering, 15*(1), 42–47. Retrieved from http://www.wiete.com.au/journals/GJEE/Publish/vol15no1/06-Saad-M-S.pdf

Saravanan, V. (2006). Sustainable employability skills for engineering professionals. *The Indian Review of World, 5*(II), 1–9. Retrieved from http://www.worldlitonline.net/sustainable-employability-skills-for-engineering-professionals.pdf

SCANS. (1991). *What work requires of schools: A SCANS Report for America 2000. The Secretary's Commission on Achieving Necessary.* Washington, D.C. Retrieved from http://scholar.google.com/scholar?hl=en&btnG=Search&q=intitle:WHAT+WORK+REQUIRES+OF+SCHOOLS+:+A+SCANS+REPORT+FOR+AMERICA+2000#0

Shah, A. A. (2009). *Contributions and limitations of self assessment of competences by higher education graduates.* Universite de Bourgogne.

Sheppard, K., Dominick, P., & Aronson, Z. (2004). Preparing engineering students for the new business paradigm of international teamwork and global orientation. *International Journal of Engineering Education, 20*(3), 475–483. Retrieved from http://citeseerx.ist.psu.edu/viewdoc/download?doi=10.1.1.162.3040&rep=rep1&type=pdf

Shuman, J., Besterfield-Sacre, M., & McGourty, J. (2005). The ABET 'professional skills'—Can they be taught? Can they be assessed? *Journal of Engineering Education, 49*(1), 41–55.

Shuman, L. J., Atman, C. J., Eschenbach, E. A., Evans, D., Felder, R. M., Imbrie, P. K., et al. (2000a). The future of engineering education. I. A vision for a New Century. *32nd Annual Frontiers in Education, 34*(1), T4A–1–T4A–15. doi:10.1109/FIE.2002.1157986

Shuman, L. J., Atman, C. J., Eschenbach, E. A., Evans, D., Felder, R. M., Imbrie, P. K., et al. (2000b). The future of engineering education III. Developing critical skills. *32nd Annual Frontiers in Education, 34*(2), T4A–1–T4A–15. doi:10.1109/FIE.2002.1157986

Skills, W., & They, W. (2009). *Transportation engineers of 2020: What skills will they need?* (pp. 1–8).

Small and Medium Enterprise (SME) Financing in Canada, Government of Canada Report (2002). Appendix A, page 100.

Spinks, N., Silburn, N., & Birchall, D. (2006). *Educating engineers for the 21st century: The industry view. Engineering*. Greenlands, UK.

Standard, U. K., & Competence, P. E. (n.d.). The accreditation of higher education programmes.

Statz, C. (2003). Two perspectives for assessing skills: what can they contribute to policy. Paper prepared for EGFSN, Sept. 2003.

Sunthonkanokpong, W. (2011). Future global visions of engineering education. *Procedia Engineering, 8*, 160–164. doi:10.1016/j.proeng.2011.03.029.

Swearengen, J., & Barnes, S. (2002). Globalization and the undergraduate manufacturing engineering curriculum. *Journal of Engineering Education, 49*(1), 255–261. Retrieved from http://onlinelibrary.wiley.com/doi/10.1002/j.2168-9830.2002.tb00700.x/abstract

Tech, V. (2011). The emergence of engineering education research as an internationally connected field of inquiry. *Journal of Engineering Education, 100*(1), 14–47.

Terenzini, P. T., & Lattuca, L. R. (n.d.). Benchmarking U.S. engineering education vis-à-vis the engineer of 2020. In *Convocation of the Professional Engineering Societies.*

Tippins, N., & Hilton, M. (2010). *A database for a changing economy: Review of the occupational information network (O* NET)*. Washington, D.C. Retrieved from http://books.google.com/books?hl=en&lr=&id=Qw1CxQ_QiV4C&oi=fnd&pg=PT5&dq=A+Database+for+a+Changing+Economy+:+Review+of+the+Occupational+Information+Network+(+O+*+NET+)&ots=Y7DnLmrMnu&sig=zCWUW91ZCn8qNneU551DIFFYmVg

Tryggvason, G., & Apelian, D. (2006). Re-engineering engineering education for the challenges of the 21st century. *IEEE Engineering Management Review*, 14–17.

UKCES. (2009). *The employability challenge full report*. London. Retrieved from www.ukces.org.uk

UNESCO. (2012). *Graduate employability in Asia* (pp. 1–96). Bangkok, Thailand.

Walther, J., & Radcliffe, D. F. (2007). The competence dilemma in engineering education: Moving beyond simple graduate attribute mapping *, *13*(1), 41–51.

Warrington, B., & Tech, M. (2010). Vision 2030: Creating the future for mechanical engineering education in the US: Engineering leadership, a need for systemic change in education and practice for a sustainable world.

WDA. (2009, August). *The Singapore employability skills system*.

Wearne, H. (1984). *Managerial skills and expertise used by samples of engineers in Britain*. Western Canada, Australia, Japan, The Netherlands and Norway. Report TMR 152, Technological Management, University of Bradford.

Wheelahan, L., Moodie, G., & Briggs, C. (2011). *Rethinking skills in vocational education and training: From competencies to capabilities*. Melbourne, Australia. Retrieved from http://www.voced.edu.au/td/nsw_68.10

Yorke, M., & Knight, P. T. (2006). *Employability embedding employability into the curriculum*. UK. ISBN: 1-905788-00-2

Zaharim, A., Yusoff, Y. M., Mohamed, A., Omar, M. Z., Muhamad, N., & Mustapha, R. (2010). Practical framework of employability skills for engineering graduate in Malaysia. In *Education Engineering Conference (EDUCON)*, 14–16 April 2010 (pp. 921–927). Madrid, Spain: IEEE. doi:10.1109/EDUCON.2010.5492478

Zaharim, A., Yusoff, Y., & Omar, M. (2009). Engineering employability skills required by employers in Asia. In *Proceedings of the 6th WSEAS International Conference on Engineering Education* (pp. 195–201). Retrieved from http://www.wseas.us/e-library/conferences/2009/rodos/EDU/EDU30.pdf

Zaharim, A., Yusoff, Y., & Omar, M. (2010). The comparison on priority engineering employability skills. *International Journal of Engineering and Technology*, *7*(2), 61–74. Retrieved from http://ijet.feiic.org/journals/2010Issue2-04-1.pdf

Zaharim, A., Yusoff, Y., Omar, M. Z., Basri, H., Nasional, U. T., Selangor, K., et al. (n.d.). Employability skills framework for engineering graduate in Malaysia. *Educational Technology*, 264–272.

MIX
Papier aus verantwortungsvollen Quellen
Paper from responsible sources
FSC® C105338

FSC
www.fsc.org

If you have any concerns about our products, you can contact us on
ProductSafety@springernature.com

In case Publisher is established outside the EU, the EU authorized representative is:
Springer Nature Customer Service Center GmbH
Europaplatz 3, 69115 Heidelberg, Germany

Printed by Libri Plureos GmbH
in Hamburg, Germany